Efficient Building Design Series

Volume III

Water and Plumbing

Ifte Choudhury
Texas A&M University

J. Trost
Texas A&M University

Prentice Hall
Upper Saddle River, New Jersey *Columbus, Ohio*

Library of Congress Cataloging-in-Publication Data
Choudhury, Ifte.
Water and plumbing / Ifte Choudhury, J. Trost.
p. cm. — (Efficient building design series : v. 3)
Includes index.
ISBN 0–13–080337–5
1. Plumbing. 2. Roof drainage. 3. Municipal water supply.
I. Trost, J. II. Title III. Series.
TH6123.C485 2000
696'.1—dc21 99–32441
CIP

Cover art © Ceri Fitzgerald
Editor: Ed Francis
Production Editor: Christine M. Buckendahl
Production Coordinator: Kathy Davis, Carlisle Publishers Services
Design Coordinator: Karrie Converse-Jones
Text Designer: Carlisle Publishers Services
Cover Designer: Ceri Fitzgerald
Production Manager: Patricia A. Tonneman
Marketing Manager: Chris Bracken

This book was set in Cochin by Carlisle Communications, Ltd. and was printed and bound by The Banta Company. The cover was printed by Phoenix Color Corp.

Pearson Education
Upper Saddle River, New Jersey 07458

Printed in the United States of America

10 9 8 7 6 5 4 3 2 1

ISBN: 0-13-080337-5

Prentice-Hall International (UK) Limited, *London*
Prentice-Hall of Australia Pty. Limited, *Sydney*
Prentice-Hall of Canada, Inc., *Toronto*
Prentice-Hall Hispanoamericana, S. A., *Mexico*
Prentice-Hall of India Private Limited, *New Delhi*
Prentice-Hall of Japan, Inc., *Tokyo*
Prentice-Hall (Singapore) Pte. Ltd., *Singapore*
Editora Prentice-Hall do Brasil, Ltda., *Rio de Janiero*

PREFACE

The snow melts on the mountain
And the water runs down to the spring,
And the spring in turbulent fountain,
With a song of youth to sing,
Runs down to the riotous river,
And the river flows to the sea,
And the water again
Goes back in rain
To the hill where it used to be.

W. R. Hearst

Water is a vital element in our life. Efficient plumbing systems make buildings livable by bringing in this vital fluid for our use and by floating the waste away. Master plumbers take pride in installations that conserve water and operate flawlessly for many years, with minimum maintenance.

This book is a primer for students, architects, constructors, managers, occupants, and owners who wish to refine and improve their understanding of building plumbing installations. It is written in a way that is easy to follow and understand. Committed readers can develop a working knowledge of the design decisions, equipment options, and operations of water supply and drainage systems of a building.

Readers who study the text and complete review problems will be able to:

- Design, size, and detail building plumbing installations.
- Select fixtures and components.
- Integrate plumbing components with site, building, foundations, structure, materials, finishes, and assemblies.

A secondary text goal is to respect the reader's time, talent, and perception by presenting material in a concise, lucid format. Illustrations are included with text to expand and reinforce the information presented, and actual building applications are emphasized for the topics covered. Study problems follow each chapter so that readers can develop confidence in their abilities to apply new knowledge and skills. The first three chapters cover general discussions about water supply, building and site drainage, site irrigation, waterscape, and methods and principles of building plumbing. Chapter 4 details plumbing installations in example residence and office occupancies, and Chapter 5 outlines plumbing work for fire and HVAC applications.

ACKNOWLEDGMENTS

I appreciate the helpful comments and suggestions from the reviewers for this volume:
D. Perry Achor, Purdue University
Glenn Goldman, School of Architecture, New Jersey Institute of Technology
Marcel E. Sammut, Architect and Structural Engineer

CONTENTS

TERMINOLOGY

acidity The ability of a water solution to neutralize an alkali or base.

activated carbon A material that has a very porous structure and is adsorbent for organic matter and certain dissolved solids.

aeration The process by which air becomes dissolved in water.

aerobic An action or a process conducted in the presence of oxygen.

air gap The unobstructed vertical distance through the free atmosphere between the lowest opening from any pipe or faucet supplying water to a tank, plumbing fixture, or other device and the flood level rim of the receptacle.

alkalinity The quantitative capacity of water to neutralize an acid. In the water industry, alkalinity is expressed in mg/L of equivalent calcium carbonate.

alum A common name for aluminum sulfate, used as a coagulant.

anaerobic An action or process conducted in the absence of oxygen.

anion An ion with a negative electrical charge.

aquifer A natural water-bearing geological formation (e.g., sand, gravel, sandstone) that is found below the surface of the earth.

area drain A receptacle designed to collect surface or storm water from an open area.

artesian Water held under pressure in porous geological formations confined by impermeable geological formations. An artesian well is free flowing.

Biochemical Oxygen Demand (BOD) The amount of oxygen (measured in mg/L) required in the oxidation of organic matter by biological action under specific standard test conditions. It is widely used to measure the amount of organic pollution in wastewater and streams.

biodegradation Decomposition of a substance into more elementary compounds by the action of microorganisms such as bacteria.

brine A strong solution of salt(s) (usually sodium chloride) with total dissolved solid concentrations in the range of 40,000 to 300,000 or more milligrams per liter.

cation A positively charged ion in an electrolyte solution, attracted to the cathode under the influence of a difference in electrical potential. Sodium ion (Na^+) is a cation.

chlorination The treatment process in which chlorine gas or a chlorine solution is added to water for disinfecting and control of microorganisms. Chlorination is also used in the oxidation of dissolved iron, manganese, and hydrogen sulfide impurities.

cleanout An accessible opening in the drainage system used for removal of obstructions.

drain A pipe, conduit, or receptacle in a building which carries liquids by gravity to waste. The term is sometimes limited to refer to disposal of liquids other than sewage.

effluent Treated wastewater discharged from sewage treatment plants.

entropy The capacity of a system or a body to hold energy that is not available for changing the temperature of the system (or body) or for doing work.

fecal matter Matter (feces) containing or derived from animal or human bodily wastes that are discharged through the anus.

flocculation The process of bringing together destabilized or coagulated particles to form larger masses or flocs (usually gelatinous in nature), which can be settled and/or filtered out of the water being treated.

flood rim The edge of a receptacle (such as a plumbing fixture) from which water will overflow.

flow rate The quantity of water or regenerant which passes a given point in a specified unit of time, often expressed in U.S. gpm (or L/min).

flush tank The chamber of the toilet in which the water is stored for rapid release to flush the toilet.

flush valve (flushometer) A self-closing valve used for flushing urinals and toilets in public buildings. This type of valve allows very high flow rates for a few seconds.

free groundwater Unconfined groundwater whose upper surface is a free water table.

grade The elevation of the invert of the bottom of a pipeline, canal, culvert, or similar conduit.

gray water Wastewater other than sewage, such as sink drainage or washing machine discharge.

groundwater Water below the surface of the ground. It is primarily the water that has seeped down from the surface by migrating through the interstitial spaces in soils and geological formations.

half-life The time required for half of the substance present at the beginning to dissipate or disintegrate.

hardness A water quality parameter that indicates the level of alkaline salts, principally calcium and magnesium, and expressed as equivalent calcium carbonate ($CaCO_3$). Hard water is commonly recognized by the increased quantities of soap, detergent, or shampoo necessary to raise lather.

head The pressure at any given point in a water system, generally expressed in pounds per square inch (psi).

hydrologic cycle The cyclic transfer of water vapor from the earth's surface through evaporation and transpiration into the atmosphere, from the atmosphere via precipitation back to earth, and through runoff into bodies of water.

interface The surface that forms a common boundary between two spaces or two parts of matter, such as the surface boundary formed between oil and water.

invert The lowest point of the channel inside a pipe, conduit, or canal.

ion exchanger A permanent insoluble material (usually a synthetic resin) which contains ions that will exchange reversibly with other ions in a surrounding solution.

makeup water Treated water added to the water loop of a boiler circuit or cooling tower to make up for the water lost by steam leaks or evaporation.

mole The molecular weight of a substance, usually expressed in grams.

Nephelometric Turbidity Unit (NTU) The standard unit of measurement used in the water analysis process to measure turbidity in a water sample.

osmosis The natural tendency for water to spontaneously pass through a semipermeable membrane separating two solutions of different concentrations (strengths).

oxidizing agent A chemical substance that gains electrons (i.e., is reduced) and brings about the oxidation of other substances in chemical oxidation and reduction (redox) reactions.

percolation Laminar gravity flow through unsaturated and saturated earth material.

permeability The ability of a material (generally an earth material) to transmit water through its pores when subjected to pressure or a difference in head. Expressed in units of volume of water per unit time per cross-sectional area of material for a given hydraulic head.

pH A measure of the relative acidity or alkalinity of water. Defined as the negative log (base 10) of the hydrogen ion concentration. Water with a pH of 7 is neutral; lower pH levels indicate increasing acidity, while pH levels above 7 indicate increasingly basic solutions.

pneumatic tank A pressurized holding tank which is part of a closed water system (such as for a household well system) and is used to create a steady flow of water and avoid water surges created by the pump kicking on and off.

potable water Water of a quality suitable for drinking.

precipitation Rain, snow, hail, dew, and frost.

runoff Drainage or flood discharge which leaves an area as surface flow or as pipeline flow, having reached a channel or pipeline by either surface or subsurface routes.

saturated zone The area below the water table where all open spaces are filled with water.

soil pipe A pipe that conveys sewage containing fecal matter.

stack A general term used for any vertical line of soil, waste, or vent pipe, except vertical vent branches that do not extend through the roof.

stack vent The extension of a soil or waste stack above the highest horizontal drain connected to the stack.

storm drain A pipe that conveys rain water, surface water, condensate, cooling water, or similar liquid wastes.

Total Dissolved Solids (TDS) The total weight of the solids that are dissolved in the water, given in ppm per unit volume of water.

venturi A tube with a narrow throat (a constriction) that increases the velocity and decreases the pressure of the liquid passing through it, creating a partial vacuum immediately after the constriction in the tube.

vortex A revolving mass of water which forms a whirlpool.

viscosity The tendency of a fluid to resist flowing due to internal forces such as the attraction of the molecules for each other (cohesion) or the friction of the molecules during flow.

waste pipe A pipe that conveys discharge from any plumbing fixture or appliance that does not contain fecal matter.

water table The upper boundary of a free groundwater body, at atmospheric pressure.

yield The amount of product water produced by a water treatment process. It is also the quantity of water (expressed in GPM, GPH, or GPD) that can be collected for a given use from surface or groundwater sources.

zeolites Hydrated sodium aluminum silicates, either naturally occurring mined products or synthetic products, with ion exchange properties.

zone of aeration The comparatively dry soil or rock located between the ground surface and the top of the water table.

Efficient Building Design Series

Volume III

Water and Plumbing

Water Supply and Wastewater

Caressing a rock
Marked with a warrior's moko
The waters of Wai-Kimihia
Warm my bones
As I sing to Hinemoa
The trees whisper
Their reply
The fantails twitter
Their reply
The stitchbirds chorus
Their reply
The water
Brings my chant
Back to me.

Paula Harris

1.0 INTRODUCTION

This chapter discusses the properties, sources, collection, and supply systems of water. Primary sources of water supply include rainfall, surface runoff, and underground reservoirs known as aquifers. Some methods of treatment of water to make it potable are discussed at length. Wastewater is generated from buildings once the supply water has been used. This is discharged from the plumbing fixtures. Methods of removal and treatment of wastewater are also discussed in this chapter.

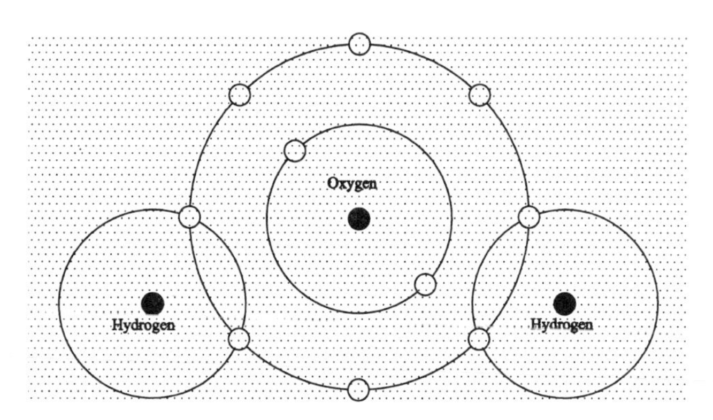

FIGURE 1.1 Chemical composition of water

1.1 WATER SUPPLY

PROPERTIES OF WATER

Water is a basic necessity of life; most of the protoplasm in the human body is comprised of water. In its purest form, water is an odorless, colorless, and tasteless liquid.

Water is a very stable chemical compound. It can withstand a temperature of about 4900°F before being decomposed into individual atoms. Its boiling point is 212°F and its freezing temperature is 32°F.

A water molecule is formed by covalent bonding between two atoms of hydrogen and one atom of oxygen. This type of bonding occurs when atoms share electrons in order to balance the structure of their outer orbits. An oxygen atom requires two electrons to complete its outer orbit while a hydrogen atom requires only one. By sharing electrons with two hydrogen atoms, oxygen can balance its outermost orbit. Water's chemical formula is, therefore, H_2O—two hydrogen atoms to one oxygen atom (Figure 1.1).

Water is a universal solvent. It acts as an ideal cleansing agent by promoting dissolution. It is also a remarkably efficient medium for transport of organic waste. The large heat-storage capacity of water makes it an effective means of heat transfer in heating, ventilating, and air-conditioning systems.

HISTORY OF WATER SUPPLY

Water has played a great role in the development of mankind throughout the history of the world. It was a determining factor in the triumph and survival of

FIGURE 1.2 A Roman aqueduct

all early human settlements. The first elaborate water supply systems were designed and constructed by the Romans (Figure 1.2). They built a system involving nine aqueducts, some of them 50 to 60 miles long, for supply of water to the city of Rome. These aqueducts had a capacity of about 84 million gallons of water per day, to be supplied mainly to the public baths, pools, and fountains.

SOURCES OF WATER

Approximately 70 per cent of the earth's surface is covered by water bodies: oceans, rivers, lakes, and glaciers. Over 97 per cent of this water is either salty (in seas and oceans) or frozen (polar regions). Out of a total of 326,000,000 cubic miles of water, only about 1,000,000 is available to us as fresh water.

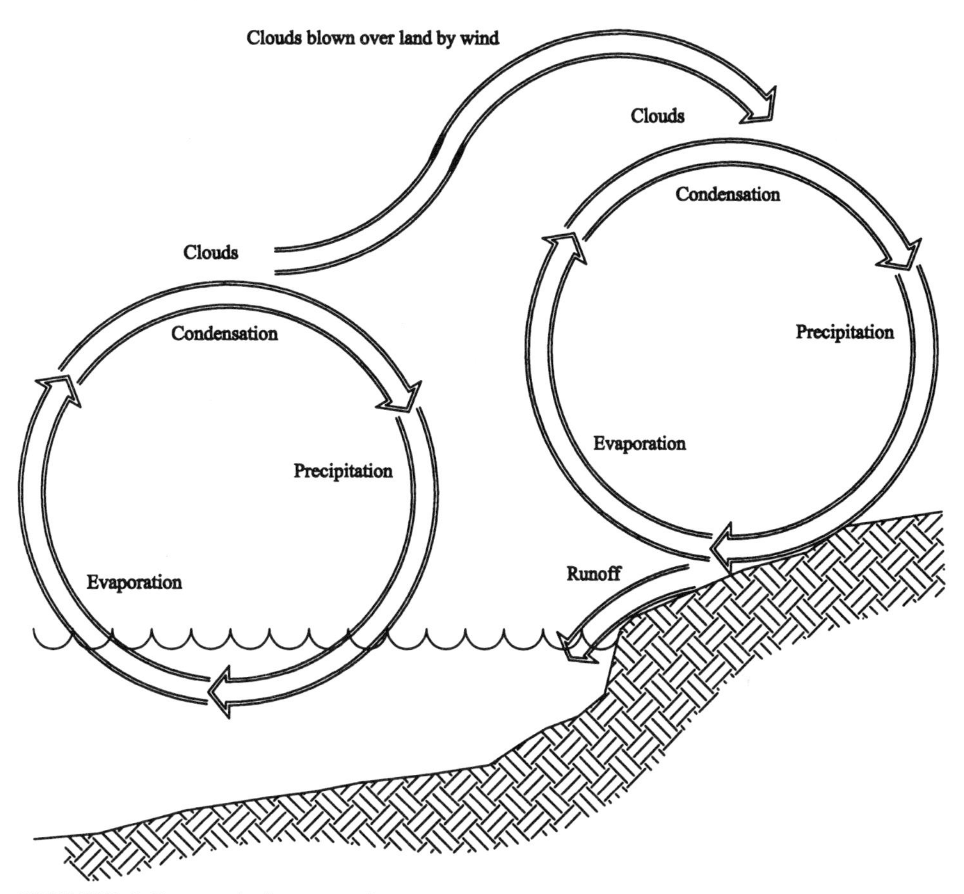

FIGURE 1.3 Hydrologic cycle

One of the primary sources of water supply is **precipitation** in the form of either rain or snowfall. Precipitation is part of the **hydrologic cycle,** a continuous exchange of water among atmosphere, the earth, and the water bodies (Figure 1.3). Water evaporates from the water bodies, soil, and vegetation, changing from a liquid to a gaseous state. The evaporated water goes up into the atmosphere, cools down, and condenses to form very small droplets of water. Clouds are formed due to the concentration of these tiny droplets. As the water vapor continues to cool and condense, larger droplets of water form around nuclei of dust or ice crystals. Finally, when the droplets coalesce, they fall to the earth as rain or snow.

The United States (excluding some desert areas) receives an average annual rainfall of 40 inches, which yields a total of about 1,080,000 gallons of water per acre. Mean annual rainfall in different regions of the United States is shown in Figure 1.4.

Rainwater is one of the purest sources of water available. Its quality almost always exceeds that of ground or surface water in terms of pollution. The pH-value of rainwater is about 5, which means that it is slightly acidic. A small amount of buffering can, however, neutralize this acid. Rainwater, therefore, can provide clean, safe, and reliable water if it is collected using a clean catchment area and then given appropriate treatment for intended uses (Figure 1.5). Table 1.1 shows quantities of water that can be obtained per square foot of catchment area for different quantities of rainfall.

Much of the rainwater that hits the ground infiltrates into the earth. Some of it is held in a shallow root zone for use by the plants. The rest of it percolates deeper into the ground through rocks and soil to form a supply of water below the earth's surface. These underground reservoirs of water are known as **aquifers.** A significant quantity of water is available from aquifers, which are geological formations that contain saturated permeable material. The upper level of an aquifer is called the water table (Figure 1.6). The impermeable rock layer below an aquifer is called an aquiclude. Water obtained from an aquifer by drilling wells is relatively pure and, therefore, requires less elaborate treatment.

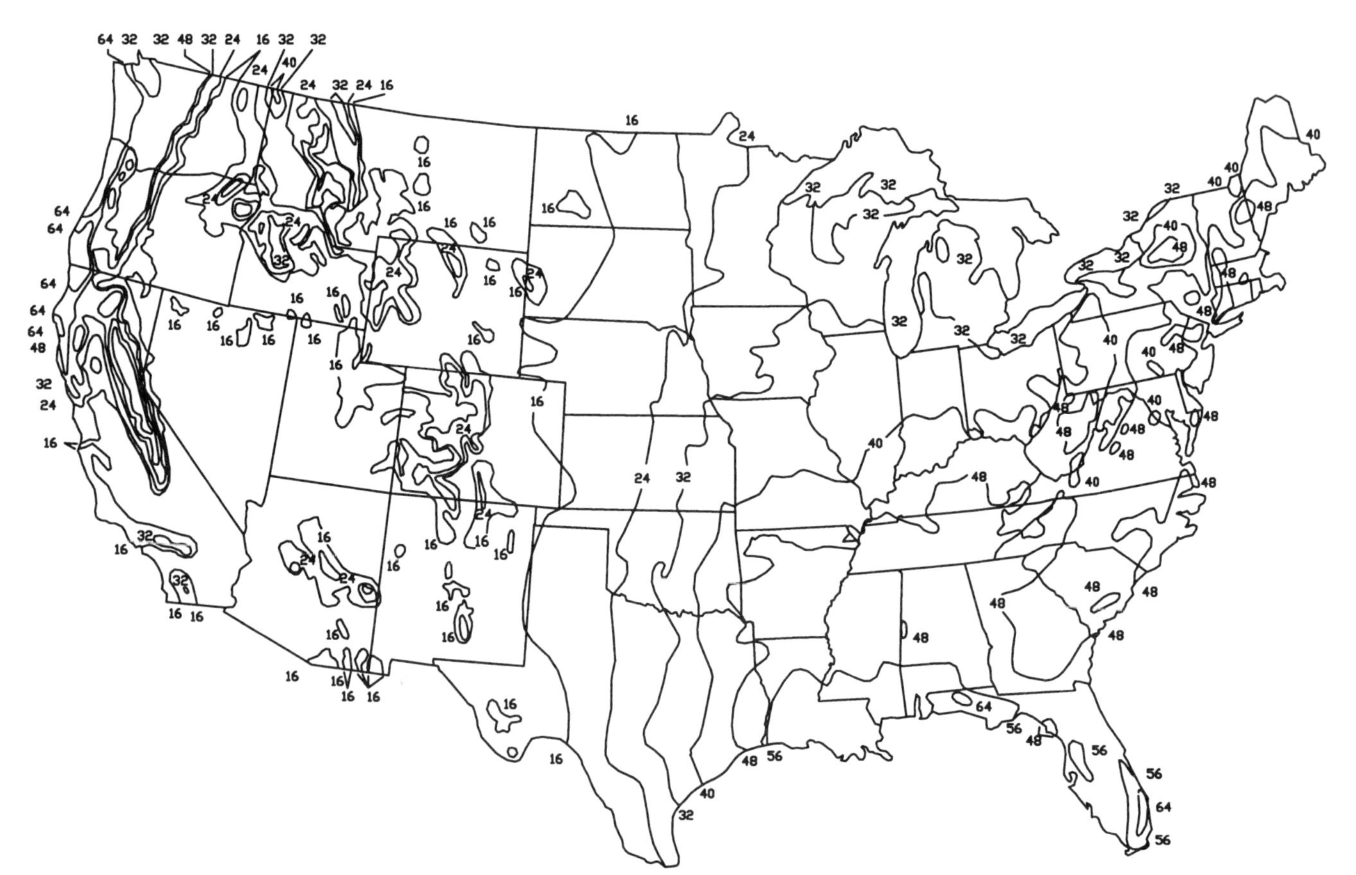

FIGURE 1.4 Mean annual rainfall (in inches) in the United States

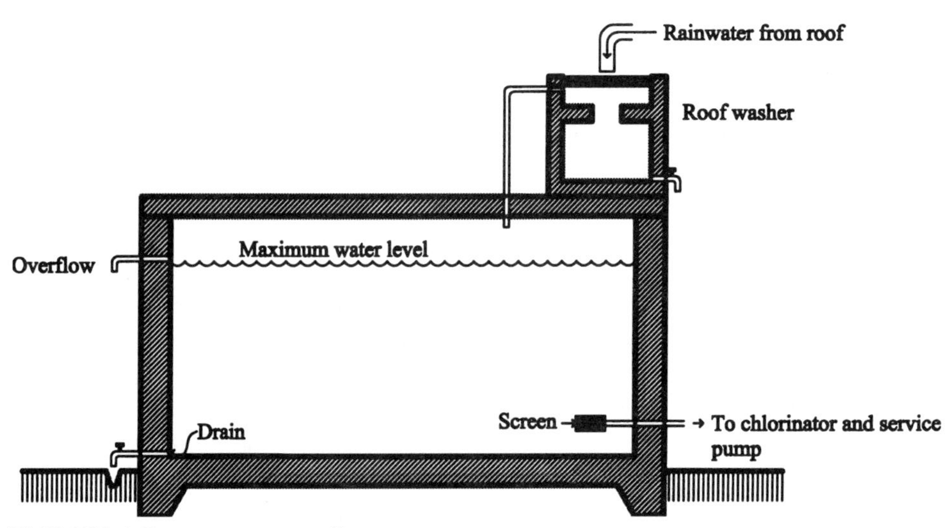

FIGURE 1.5 Rainwater collection

TABLE 1.1

Quantity of Rainwater Provided Per Square Foot of Catchment Area

Mean Annual Rainfall (In Inches)	*Water Available Per Square Foot of Catchment Area (In Gallons)*
10	4.20
15	6.20
20	8.30
25	10.40
30	12.50
35	14.50
40	16.60
45	18.60
50	20.80
55	22.90
60	25.00

Water that does not infiltrate to the ground is known as surface **runoff.** It flows across the ground and is collected in rivers, streams, lakes, or other depressions. These bodies of water are another source of water supply. Surface water becomes quickly contaminated. Streams and rivers transport silt and clay along with water, making the water turbid. Decaying algae and other organic matter add color and odor to water. Biological contaminants such as protozoa and bacteria occur naturally in surface water.

Treatment of surface water is extensive, which increases the cost of such water. Surface water, however, is preferable in regions where pumping of groundwater may cause subsidence. Water from seas and oceans contain a very high quantity of **"total dissolved solids" (TDS).** It is about 35,000 ppm, which is 70 times higher than the maximum allowable quantity for continual human consumption. (TDS in rainwater ranges from 100 to 150 ppm.) If water from such sources is to be used for consumption, then the TDS must be removed by distillation, electrodialysis, or ion exchange methods. Seawater reverse osmosis (SWRO) plants may also be constructed for the purpose. Such methods, however, are quite expensive.

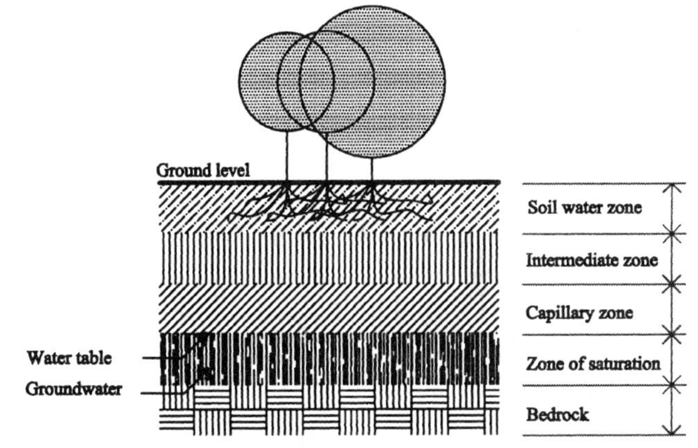

FIGURE 1.6 Water table

Cost per 1000 gallons of water desalinized using reverse osmosis is about $4.00 ($5.50 at 1997 prices).

Low-cost systems, utilizing solar energy, can be designed for desalination of brackish water. A solar still consists of a basin with a sloping glass cover and low walls, resembling a house with a glass roof (Figure 1.7). The base is made of blackened fiberglass with insulation at the bottom. Gutters run on the inside, at the bottom edge of the glass. Salty

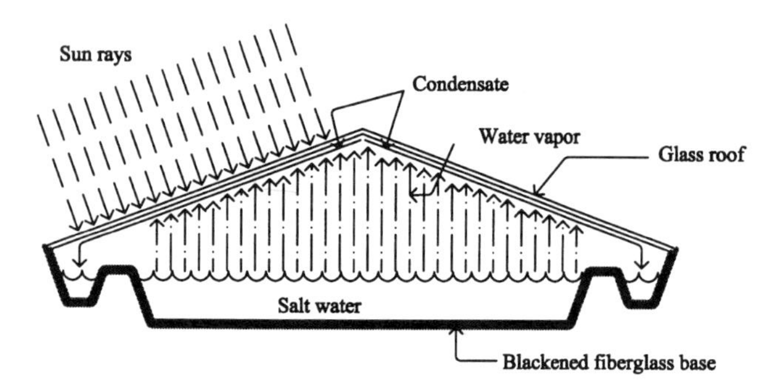

FIGURE 1.7 Solar distillation

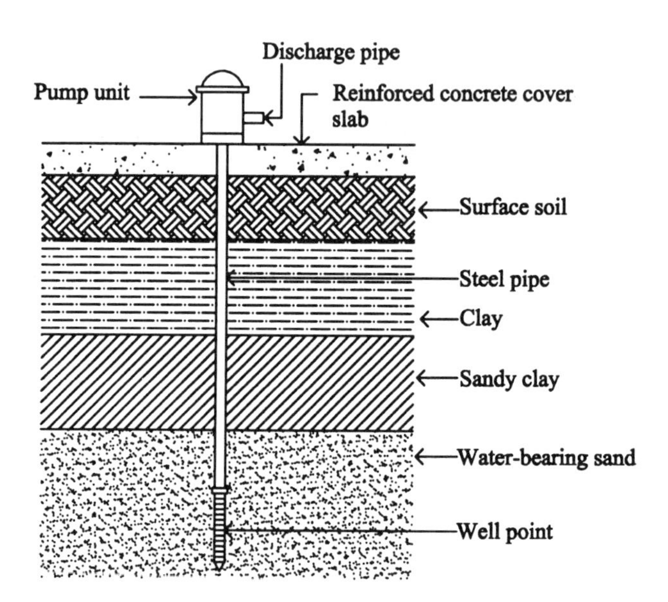

FIGURE 1.8 Driven well

water, about 12 to 16 inches deep, is introduced in the basin and exposed to the sun. As the water evaporates, the air inside the basin becomes saturated, water condenses on the glass surface, flows down the inside of the glass, and collects in the gutter. The water is then directed to a storage tank.

GROUNDWATER SUPPLY SYSTEM

Groundwater is usually collected by digging either shallow or deep wells. When the depth of the well is less than 25 feet, it is called a shallow well. Such wells can be dug, driven, or drilled. Wells deeper than 25 feet are called deep wells. They are either bored or drilled through the earth and rocks.

Dug wells are quickly contaminated by surface-water flow or by seepage of polluted groundwater. Being generally shallow, these wells tend to fail during periods of drought or whenever water is rapidly withdrawn.

Driven wells are very simple and inexpensive to construct. Such wells are driven into the ground using a well point attached to a steel pipe. The well point, having a diameter of 1¼ to 2 inches, is also made of steel (Figure 1.8).

Bored wells are similar to dug wells. They are dug using earth augers to a depth not exceeding 100 feet. These wells are usually lined with vitrified tile, concrete, or metal. The diameter of a bored well may range from 2 to 30 inches (Figure 1.9).

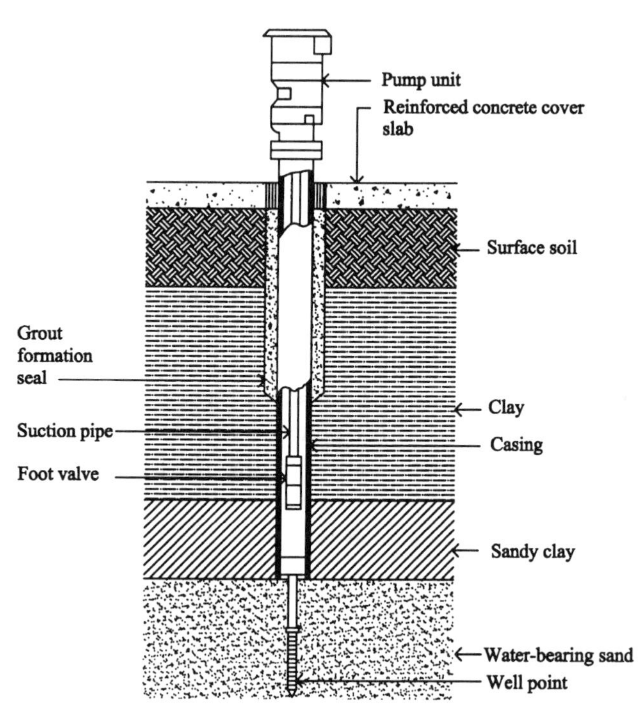

FIGURE 1.9 Bored well

Drilled wells are very deep, sometimes in excess of 1000 feet (Figure 1.10). They are constructed using machine-operated drilling equipment or rigs. Techniques employed for the construction of drilled wells are percussion and rotary. Percussion drilling utilizes a heavy drill bit and a stem. These elements are alternately raised and lowered to pulverize the earth. Water is mixed with the drilled earth to form slurry that is periodically removed to the surface.

Rotary drilling methods utilize a cutting bit at the end of a drill pipe, a revolving table for drill pipe passage, a drilling fluid or pressurized air, and a power

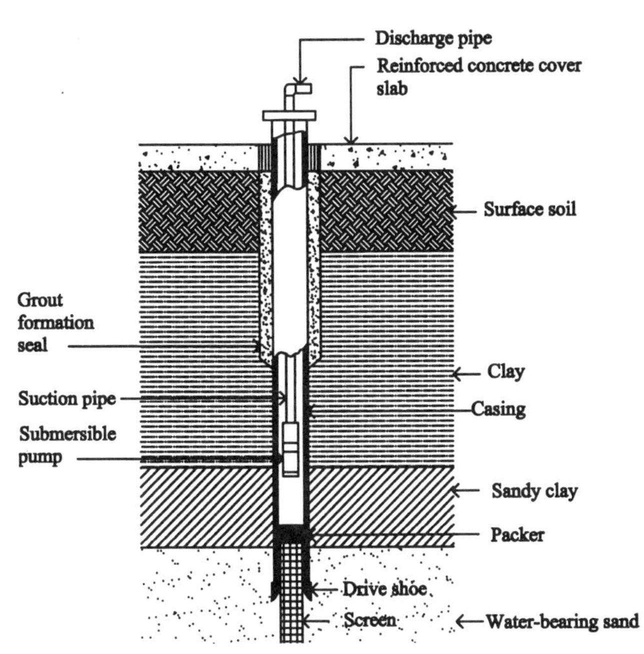

FIGURE 1.10 Drilled well

source to drive the drill. As drilling proceeds, the cutting bit rotates and advances to break up the rock formation. Drilling fluid is continuously pumped down the drill pipe for removal of cuttings to the surface. A metal case is installed, and sometimes grouted in place, after the drill stem is withdrawn.

PUMPS

Two categories of pumps are used for collection of water from wells: positive displacement and dynamic. In positive displacement pumps, energy is periodically added by application of force to the movable parts of a pump to increase the water velocities. In dynamic pumps, energy is continuously added to increase the water velocities within the machine.

There are two major classes of positive displacement pumps: reciprocating and rotary. A reciprocating pump works on a two-stroke principle (Figure 1.11). It consists of a piston within a cylinder, which draws in water on an ingoing stroke and delivers the same on the outgoing stroke. It is equipped with check valves (one-way valves) on both the suction and delivery sides.

A rotary pump has a helical or spiral rotor within the pumping chamber that rotates during the operation of the pump. Water entering the suction side is forced to the delivery side due to the rotation of the rotor (Figure 1.12).

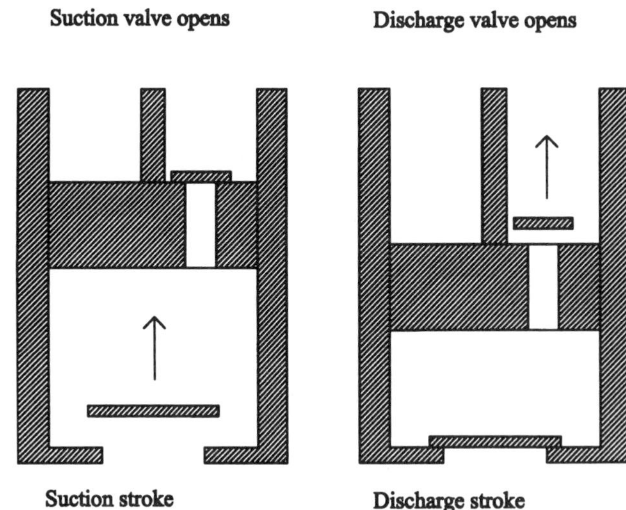

FIGURE 1.11 Operation of a reciprocating pump

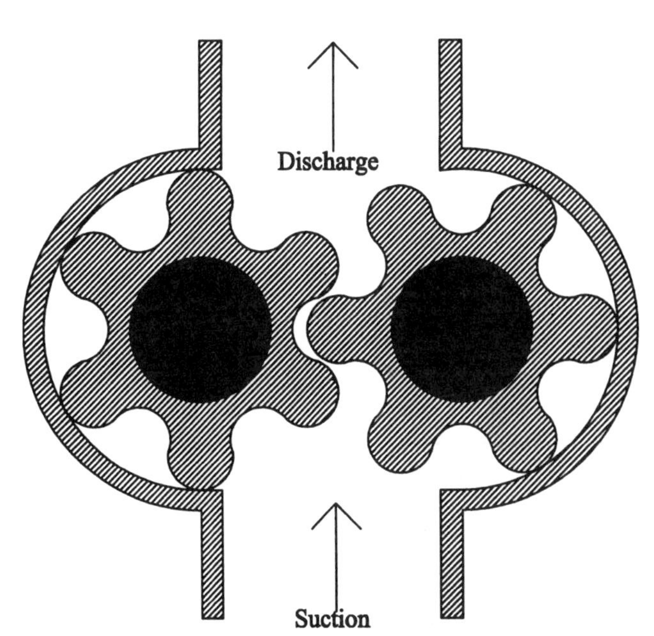

FIGURE 1.12 Operation of a rotary pump

The major classes of dynamic pumps used for water supply are centrifugal and jet pumps. Centrifugal pump is a general description for equipment that utilizes an impeller mounted on a rotating shaft for suction and delivery of water. An impeller together with its casing is called a stage. Pumps with more than one stage are called multistage pumps. Two common types of centrifugal pumps that are used for collecting well water are turbine and submersible. The assembly of a turbine pump consists of a suction head, an impeller or a set of impellers, discharge bowl, intermediate bowl or bowls, and a vertical impeller

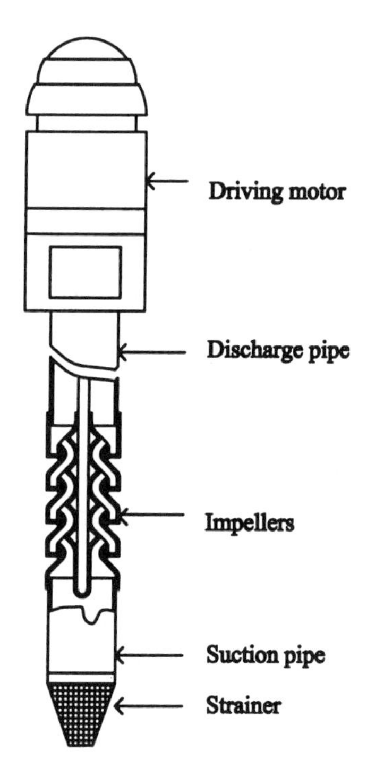

FIGURE 1.13 Turbine pump

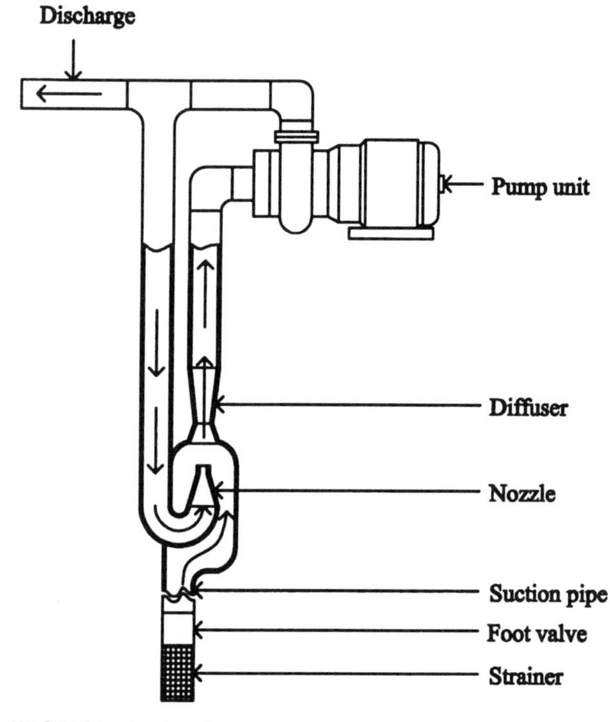

FIGURE 1.14 Jet pump

shaft along with various bearings (Figure 1.13). The driving motor of the pump is located over well casing at grade.

Pump and motor in a submersible pump are coupled as one unit. The unit is completely submerged in the well along with the impellers. The need for lengthy pump shaft is thus eliminated.

A jet or ejector pump is a special-effect dynamic pump that consists of a nozzle, a diffuser, and a suction chamber. It utilizes the motive power of a high-pressure stream of water directed through a nozzle designed to produce high velocity. The high-velocity jet stream creates a low-pressure area in the mixing chamber that causes the suction water to flow into this chamber. The diffuser entrains and mixes the water and converts velocity to pressure energy. The increased pressure pushes the suction water toward the discharge port (Figure 1.14).

The vertical distance traversed by water from the point of intake to the level where it is supplied is called the head of a pump. It is dependent on the total pump pressure that includes suction lift, static head, and friction loss plus pressure head. The relationship between pressure (P) and head (H) can be expressed as follows:

$$\begin{aligned} P\ (\text{lb/in}^2) &= [62.4\ (\text{lb/ft}^3) \cdot H\ (\text{feet})]/144 \\ &= H\ (\text{feet})/2.31 \end{aligned} \tag{1}$$

WATER QUANTITY

The supply water is usually estimated in terms of gallons per person per day (gppd). Domestic water consumption can be broken down into two categories: indoor use and outdoor use. An average residential water consumption pattern in the United States is given in Table 1.2.

WATER QUALITY

Water used for consumption (bathing, cooking, drinking, etc.) should be **potable,** i.e., safe enough to drink. It also should meet the minimum level of other qualities based on several major characteristics, as follows:

Turbidity The clarity of water is measured by the presence of insoluble suspended solids, i.e., its turbidity. Turbidity is more common in surface than

TABLE 1.2

Residential Water Consumption Pattern in the United States

Type of Use		Quantity of Water (In gppd)	
Indoor	Bathing & personal hygiene	21	
	Flushing toilets	30	
	Laundry & dishwashing	15	
	Drinking & cooking	4	
	Total indoor		70
Outdoor	Yard irrigation, car wash, etc.		70
Total			140

groundwater. Clarity of water is determined by the amount of light scattered by solid particles in a sample of water. Potable water should not contain more than 5 turbidity units (TU). One turbidity unit equals 1 mg of suspended matter per liter of water sample.

pH-value **pH** refers to "potential hydrogen," and is a measure of hydrogen ion concentration in water. It is expressed as follows:

$$\text{pH} = \log(1/H^{+}) \qquad (2)$$

where H^{+} = number of hydrogen ions.

The higher the concentration of hydrogen ions in water, the lower is the pH-value, indicating a high concentration of acid. Addition of alkali reduces the number of free hydrogen ions, causing an increase in pH-value. It is measured on a scale ranging from 0 to 14. The pH-value is 0 when water is extremely **acidic;** the value is 14 when water is extremely **alkaline.** At a pH-value of 7, water is considered to be neutral. Recommended pH-value of water ranges from 6.5 to 8.5.

Hardness It is a measure of detergent-neutralizing ions present in water. A **hardness** of about 300 mg/L in public water supplies is considered to be excessive. It results in hindering the cleansing action of detergents and formation of scales in cooking utensils and supply pipes. Calcium and magnesium salts predominantly cause hardness. A level of hardness between 60 and 120 mg/L is acceptable in potable water.

Biochemical Oxygen Demand (BOD) It is the amount of oxygen, measured in mg/L, required in the oxidation of organic matter in water by biological action. **BOD** is thus a measure of pollution in wastewater and streams. Supply water should not contain more than 4 mg/L of BOD.

WATER TREATMENT

A supply system should provide potable water that is safe from chemical and bacteriological points of view. Domestic water should also be free from unpleasant taste and smell, and improved for human health. Water should be conditioned by utilizing different treatment processes before it is supplied.

Oxidation This is a method of breaking down organic pollutants or chemicals present in supply water by addition of oxygen. This process also helps remove such minerals as iron and manganese, thereby improving the taste and color of water.

Sedimentation This process involves the removal of suspended solids from water by allowing the particles to settle out. The water has to be kept still in a pond or a basin for at least 24 hours for sedimentation to take place. This method helps to remove the turbidity of water. Adding a chemical such as hydrated aluminum sulfate (commonly known as **alum**) can increase the efficiency of the method. This chemical causes fine suspended particles in water to coagulate and settle down to the bottom of the sedimentation pond.

Filtration This is a process of removing suspended matter from water by passing it through beds of porous materials such as sand, gravel, porous stone, diatomaceous earth, activated charcoal, or filter paper. Filtration not only removes suspended matter but also helps remove some bacteria and improves turbidity, color, taste, and odor of water to a certain extent.

Disinfecting Disinfecting may be a simple or complex matter depending on the source of water supply. Chlorine is extensively used as a disinfecting agent to treat water for municipal and individual supplies. Chlorine has a strong oxidizing power on the chemical structure of bacterial cells; it destroys the enzymatic processes of the cells required for their existence.

Ozone is also used as a disinfectant. The use of ozone helps remove color, taste, and odor that may be present in water. It, however, does not provide a lasting residual like chlorine in treated water. Therefore, a secondary **chlorination** is recommended in order to provide a protective residual in the distribution system.

Softening Softening is the process of removal of hardness from water. Hardness occurs when calcium and magnesium carbonates dissolve in ground water. The process begins when rainwater dissolves carbon dioxide to form carbonic acid:

$$H_2O + CO_2 \Rightarrow H_2CO_3 \qquad (3)$$

This weakly acidic water dissolves limestone when it comes into contact with the material and forms solutions of calcium and/or magnesium bicarbonate as follows:

$$H_2CO_3 + CaCO_3 \Rightarrow Ca(HCO_3)_2 \qquad (4)$$

$$H_2CO_3 + MgCO_3 \Rightarrow Mg(HCO_3)_2 \qquad (5)$$

The hardness may be removed utilizing various methods of water softening. One of the most popular methods is the **zeolite** system. The process involves an exchange of calcium or magnesium ions with sodium ions. The softener consists of a resin tank filled with ion-exchange resins, a group of hydrated sodium aluminosilicates (Figure 1.15). Another component of the system is a brine tank containing sodium solution for regenerating the resins. When water is passed through the sodium-saturated resins, they remove calcium and magnesium ions from water and release sodium ions into it. Hardness of water is thus removed.

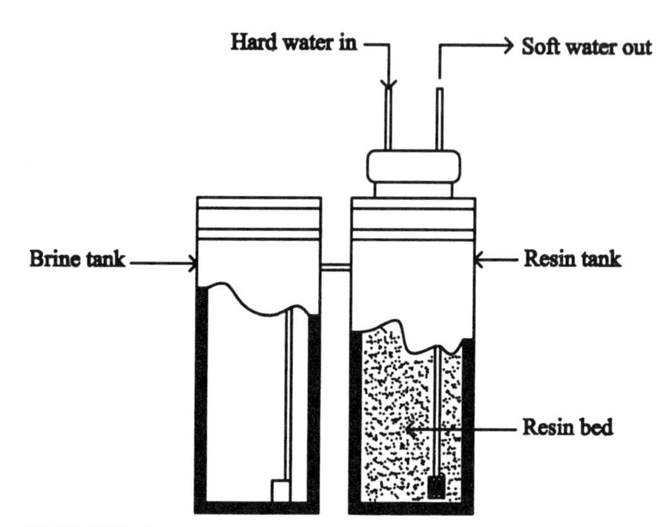

FIGURE 1.15 Water softener

Use of a deionizing system can also provide water softening. This method is capable of removing hardness caused by not only calcium and magnesium ions but also by other types of alkali metal ions.

1.2 WASTEWATER

SEWER

Wastewater generated from buildings has to be transported by sewers either to central facilities for treatment and disposal, or may be treated using individual waste systems. Drainage in buildings has two major components: sanitary waste and storm water. Sanitary waste consists of liquid discharged from plumbing fixtures. A pipe that carries this waste is called sanitary sewer. Storm water consists of rainwater, surface water, condensate, cooling water, or other similar liquid wastes. A **drain** that carries this type of waste is called a **storm sewer.** Since sanitary waste contains a high level of pollutants and BOD, it is necessary that sanitary and storm sewers should be separate.

MUNICIPAL SEWAGE TREATMENT

Household sewage is largely composed of water with very small amounts of solid material. The quantity of both suspended and dissolved solids in waste is only about 0.1 percent of the total volume.

Sanitary sewers in urban areas carry sanitary waste to a central treatment plant before it is discharged into streams or rivers. The treatment is provided in four stages:

Sedimentation The waste is first diverted to a grit chamber where the larger solids and heavy objects settle out. It then moves on to a sedimentation tank where lighter suspended materials settle to the bottom and grease floats on top (Figure 1.16). The mass of sediments is called sludge.

The sludge and scum from the sedimentation tank is usually pumped to a sludge thickener for hydration. The thickened sludge then may be transferred to a digester where it is anaerobically decomposed. Methane gas produced during the process is a good source of energy for operating the treatment plant. The decomposed sludge from the digester may be used as a fertilizer.

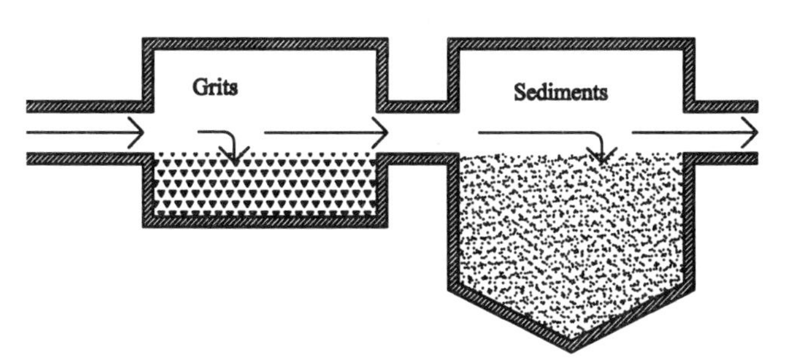

FIGURE 1.16 Sedimentation tank

Aeration The waste is pumped to an **aeration** tank where it mixes with oxygen in air for several hours (Figure 1.17). During this time, colonies of **aerobic** bacteria break down the organic matter present in waste. Aeration of the sewage results in the removal of about 90 percent of organic matter.

Chlorination The **effluent** from the aeration tank is carried to a chlorination chamber where it is treated using chlorine gas for about 15 to 30 minutes. This process kills the remaining organisms in the effluent.

Dechlorination After the effluent has been disinfected, it is given a further treatment to remove the chlorine. The process involves either the use of some chemicals such as sodium metabisulfate, or the use of oxygen in air. Once chlorine is removed, the effluent is released into streams or rivers, or is used for irrigation. It has to be ensured that the effluent coming out from the final stage of treatment contains a low BOD.

SEPTIC TANK SYSTEMS

A septic tank is an individual waste system for households in areas that are not connected to a municipal treatment system. It consists of a tank for primary treatment of waste and a method of filtration of the effluent as secondary treatment.

The system works by allowing wastewater to separate into layers and decomposing the organic matter while it is contained within the tank. **Anaerobic**

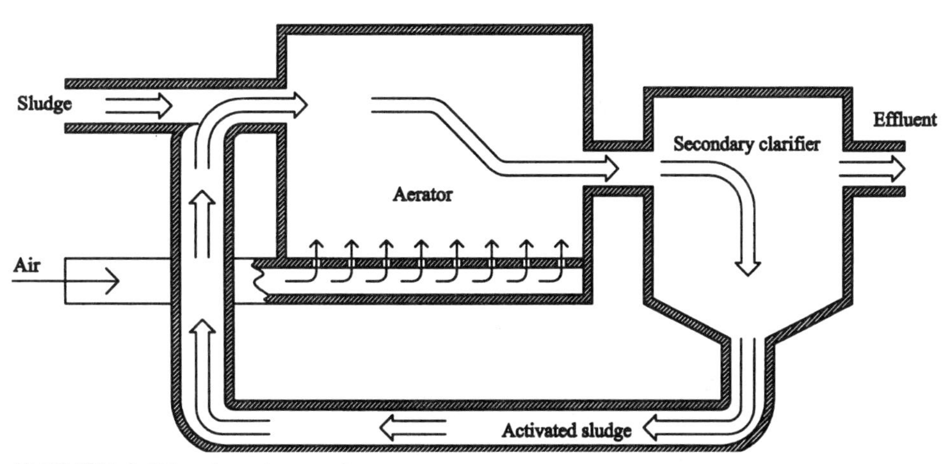

FIGURE 1.17 Aeration tank

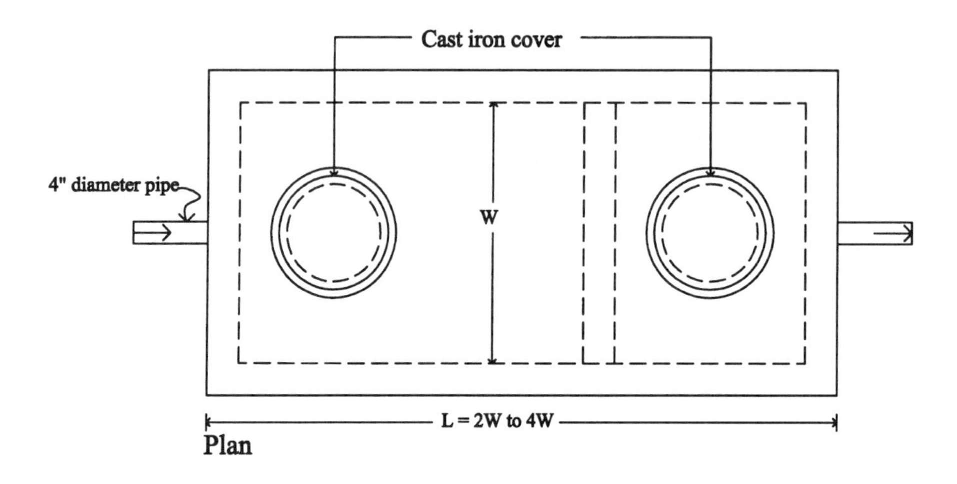

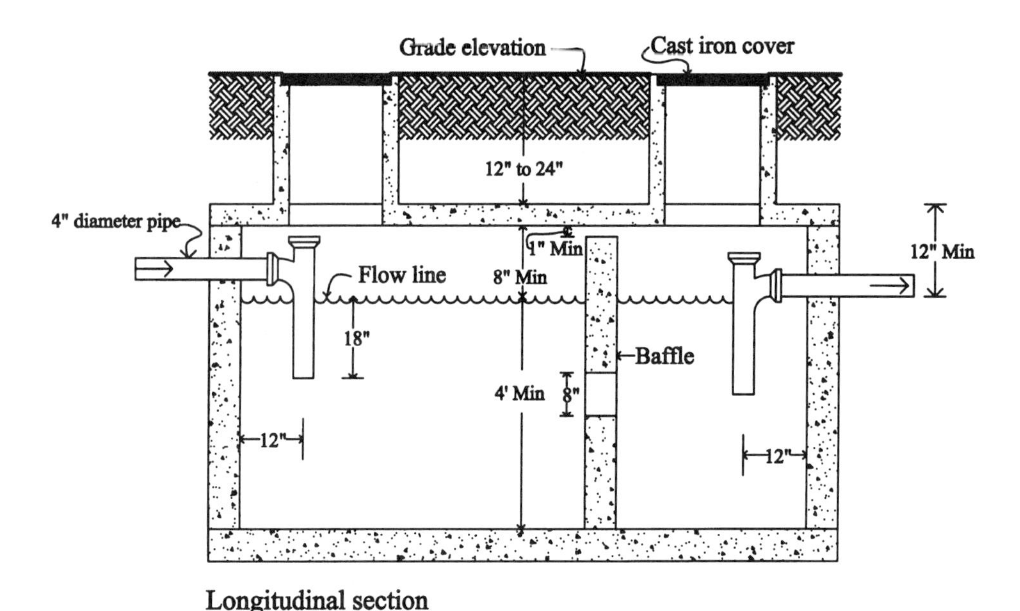

FIGURE 1.18 Rectangular septic tank

bacteria, naturally present in the septic systems, digest the solids that settle out at the bottom of the tank. The outflow, or effluent, from the tank is then given a secondary treatment.

Construction of Septic Systems Septic tanks are usually constructed underground using brick, stone, concrete blocks, precast concrete, fiberglass, or steel. Use of precast concrete, however, is very common.

The tank may either be rectangular or circular in configuration (Figure 1.18). It is divided into two chambers using a baffle wall with an opening 18 inches below the flow line of waste. A 4-inch-diameter house sewer delivers the waste into the first chamber where the grits and organic solids settle out at the bottom. Grease and other lighter material rise to top forming scum. The baffle wall stops the sludge and scum from entering the second chamber. The finer solids that enter the second chamber along with the liquid settle at the bottom of this chamber. The sludge retained in the chambers are decomposed due to the action of anaerobic bacteria, while the clarified effluent is discharged either to a seepage pit or a drainfield for secondary treatment. Septic tank capacities needed for different

TABLE 1.3

Septic Tank Capacity

Single Family Dwellings: Number of Bedrooms	*Multiple Dwelling Units or Apartments: One Bedroom Each*	*Other Uses: Maximum Fixture Units Served*	*Minimum Septic Tank Capacity (In Gallons)*
1–3		20	1000
4	2	25	1200
5 or 6	3	33	1500
	4	45	2000
	5	55	2250
	6	60	2500
	7	70	2750
	8	80	3000
	9	90	3250
	10	100	3500

TABLE 1.4

Septic Tank Sizes

Capacity (In Gallons)	*Length*	*Width*	*Air Space*	*Liquid Depth*
1000	8′-0″	4′-0″	1′-0″	4′-0″
1200	9′-0″	4′-6″	1′-0″	4′-0″
1500	9′-6″	4′-9″	1′-0″	4′-6″
2000	10′-6″	5′-3″	1′-3″	4′-9″
2250	11′-0″	5′-6″	1′-3″	5′-0″
2500	11′-6″	5′-9″	1′-3″	5′-0″
2750	12′-0″	6′-0″	1′-3″	5′-3″
3000	12′-6″	6′-3″	1′-3″	5′-3″
3250	13′-0″	6′-6″	1′-3″	5′-3″
3500	13′-6″	6′-9″	1′-3″	5′-3″

buildings and representative tank dimensions are given in Tables 1.3 and 1.4, respectively.

Seepage Pit A seepage pit or leeching pool is also commonly constructed using precast concrete (Figure 1.19). The pit has a circular wall with perforations that allow the effluent to seep into the gravel surrounding it before being absorbed by the soil. A seepage pit cannot be constructed in a place with high water table. The bottom of the pit should be placed at least 2 feet above the water table.

Drainfield A drainfield is an area where the effluent from a septic tank is distributed by horizontal underground piping in order to expedite the process of natural leeching and percolation through the soil. The drainage system consists of 4-inch-diameter clay tiles or perforated plastic pipes buried in trenches

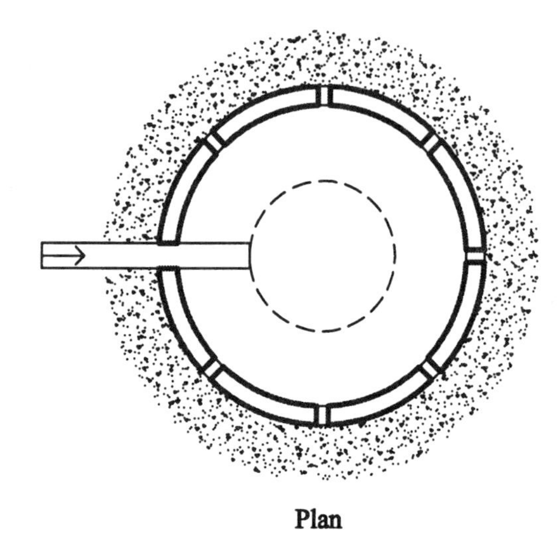

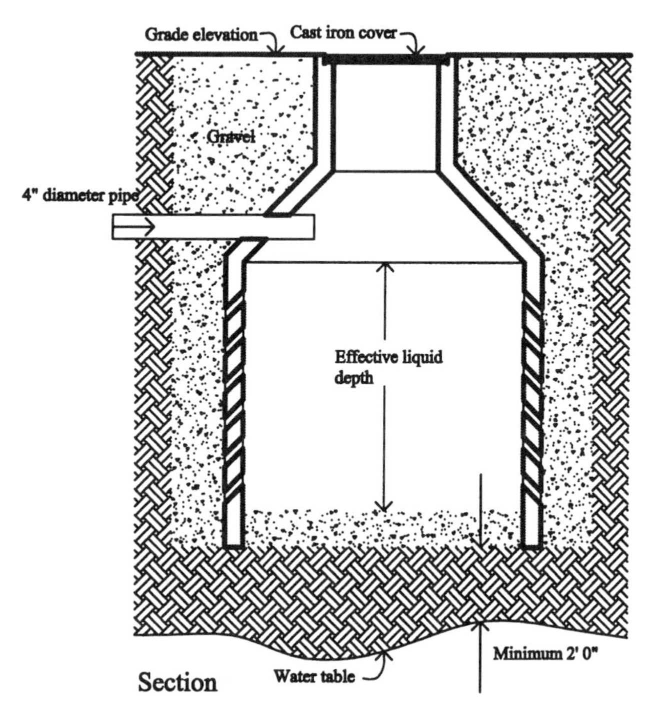

FIGURE 1.19 Seepage pit

(Figure 1.20). Clay tiles are laid with a separation of ¼ inch between the ends to allow the effluent to run out. The tiles or pipes are covered on all sides by clean stone or gravel. The tops of the trenches are backfilled with earth. Drainfield areas required for residential building are given in Table 1.5.

Percolation Test It is necessary to determine the absorption capacities of soil before the construction of a drainfield. This can be done using a method known as percolation test. In order to perform percolation tests, several straight-sided holes, each at least 4 inches in diameter, are dug down to the level where drainfield pipes are to be laid. These holes should be at least 14 inches in depth. Smeared soil surfaces are then removed from the holes and 2 inches of coarse sand or fine gravel are placed at the bottom. The purpose of this filling is to protect the holes from scouring. They are now filled with clear water to a minimum depth of 12 inches. This water level is maintained for at least 4 hours for all types of soils except sands.

After soaking, 6 inches of water is added over the sand or gravel filling. The drop in this water level is measured from a fixed reference point at ground level, at approximately 10-minute intervals for one hour. The drop that occurs during the final 30-minute period is used for calculating the absorption rate. This rate is expressed in number of minutes required for water to drop an inch.

LAGOONS

Sewage lagoons are shallow ponds constructed to hold sewage while aerobic bacteria decompose the waste. Lagoon floors and sides should be relatively impervious, to minimize seepage and contamination of groundwater. A lagoon is required to provide a 7-day retention of sewage. Soils that contain a large amount of organic matter are not suitable for the floor of an aerobic lagoon.

STORM WATER

Storm sewers, designed commonly as gravity flow conduits, usually carry storm water. Infiltration of storm water into sanitary sewers should be avoided as far as possible. Excessive clean water inputs in sanitary sewer result in overloading the treatment plants.

Storm water should be disposed of in such a manner that it does not create damages to property or people during periods of heavy rainfall and subsequent runoff. Surface drainage has to be considered as a function of grading.

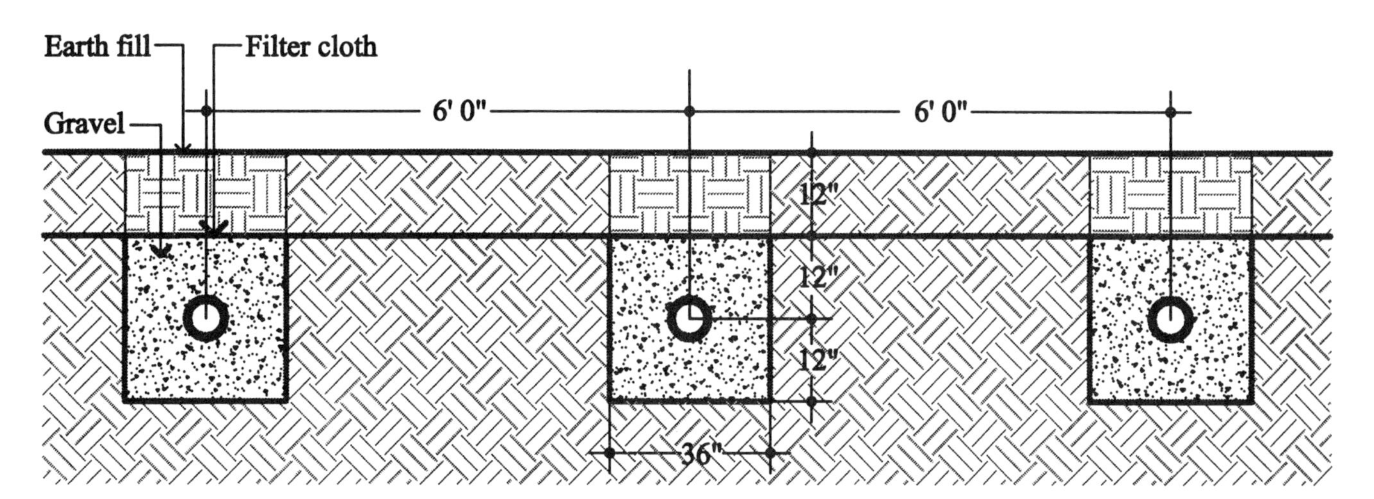

Cross section

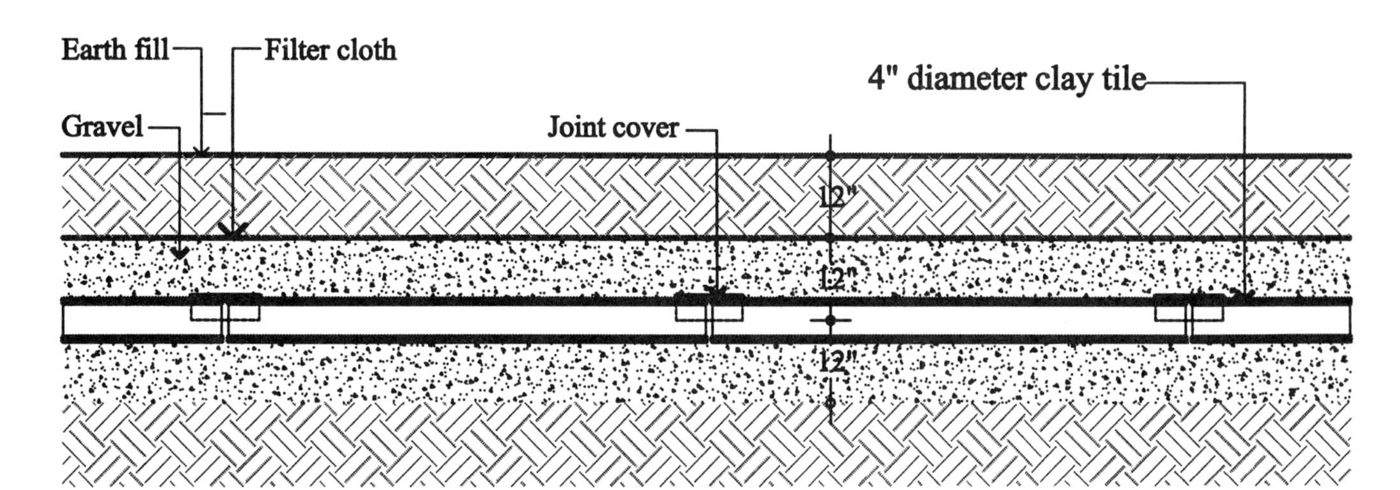

Longitudinal section

FIGURE 1.20 Drainfield trench

TABLE 1.5

Drainfield Area Per Bedroom

Average Percolation Rate (Minutes/Inch)	*Area of Trench Per Bedroom (Sq. Feet)*	*Length of Trench in Feet*		
		18″ Wide Trench	*24″ Wide Trench*	*36″ Wide Trench*
5	125	84	63	42
10	165	110	83	55
15	190	127	95	64
20	215	144	108	72
30	250	167	125	84
45	300	200	150	100
50	315	210	158	105
60	340	227	170	113
70	360	240	180	120
80	380	254	190	127
90	400	267	200	134

REVIEW QUESTIONS

1. What is the approximate temperature at which a water molecule is decomposed into individual atoms of hydrogen and oxygen?
2. What is surface runoff?
3. What is the maximum recommended limit of turbidity units in potable water?
4. What is sedimentation?
5. Hardness in water is predominantly caused by the presence of _____ .
6. What is indicated by the pH-value of water?
7. What is a drainfield?
8. What is the method used to determine absorption capacities of soil?
9. What is a storm sewer?
10. Name the term used to indicate the height up to which water can be lifted by a pump from the point of intake to the level of supply.

ANSWERS

1. About 4,900°F.
2. Water that does not infiltrate into the ground.
3. 5
4. It is the process of removing suspended materials from water.
5. Calcium and magnesium ions.
6. pH-value is a measure of hydrogen ion concentration in water. A low pH-value indicates a high concentration of hydrogen ions and high acidity.
7. It is a horizontal area used for leeching and percolation of effluent from a septic tank through the soil.
8. Percolation test
9. It is the drainpipe that carries rainwater, surface water, condensate from HVAC systems, and other similar liquid wastes.
10. Head of a pump

CHAPTER

2

Site and Roof Water Drainage

Marvels are there in every land,
Where nature shows her wondrous hand;
Rivers are there in East and West,
Where tourists throng for change and rest.

J. A. O'Reilly

2.0 INTRODUCTION

This chapter discusses the principles of site and roof drainage. Proper site planning and development is essential for efficient site drainage. Capably designed, constructed, and maintained drainage systems, both for the site and roof, promote the comfort and durability of a building. Site development includes landscaping. The chapter elaborates on the water requirement for maintenance of landscaping and techniques of irrigation. It also deals with the design of parking areas as part of site development.

Apart from practical purposes, water may also be used to enhance the aesthetic qualities of a site. Issues related to the design of pools and fountains are discussed in this chapter.

2.1 SITE DRAINAGE

OBJECTIVE OF SITE DRAINAGE

The principal objective of site drainage is to keep storm water from entering the buildings located on the site. Storm water should be disposed of in such a way that it does not create any damage either to the properties or people during periods of heavy rainfall and subsequent runoff.

SUBSURFACE SYSTEMS

Subsurface systems are the primary techniques of removing rainwater from site. The basic objectives of subsurface systems are collection, transfer, and disposal of surface runoff. In order to ensure that water does not enter the buildings, it is necessary that they be positioned at a higher elevation than the street level. Proper grading is an integral part of positive site drainage. Surface drainage has to be considered as a function of grading. Storm water should be diverted from the buildings by grading the site to provide at least six inches of vertical fall in the first ten feet of horizontal distance toward the street level. It should be disposed of in such a manner that it does not create damages to property or people during periods of heavy rainfall and subsequent runoff. Since water always runs downhill and perpendicular to the contour line, it may be necessary to revise the site contours for easy removal of surface water. See Figures 2.1(a) and 2.1(b).

Drainage swales are required in order to maintain the flow of surface water between buildings. A swale is a shallow ditch that provides a channel for flow of water to catch basin or other outlet. Collection of surface runoff is done by **area drains,** catch basins, and trench drains. See Figures 2.2(a), 2.2(b), and 2.2(c).

Erosion control is also an integral part of any drainage plan. Surface runoff can create substantial erosion of the site. Erosion should be controlled in order to maintain an effective and clear drainage system with a minimum of maintenance. This can be achieved by adopting either mechanical or vegetative measures or both. Mechanical measures include proper grading of site, creating diversions, or constructing sediment basins. Vegetative measures include preserving trees and shrubs on site, planting of turf or ground cover immediately after grading, using straw mulch to protect constructed slopes, and using fibrous materials directly on soil to protect newly seeded channels.

Once the water is collected, it is carried by storm sewers and disposed of in streams or rivers. Storm sewers, designed commonly as gravity flow conduits, usually carry storm water. Infiltration of storm water into sanitary sewers should be avoided as far as possible. Excessive clean water inputs in sanitary sewer result in overloading the treatment plants.

GUIDELINES FOR SURFACE DRAINAGE

- Runoff should not be purposefully redirected from its natural course on one's own property such that it creates a problem on another property.
- A method of reducing the velocity of surface runoff should always be considered so that it might be absorbed in the soil.
- A drainage plan and grading should be so devised as to take advantage of the existing natural systems.

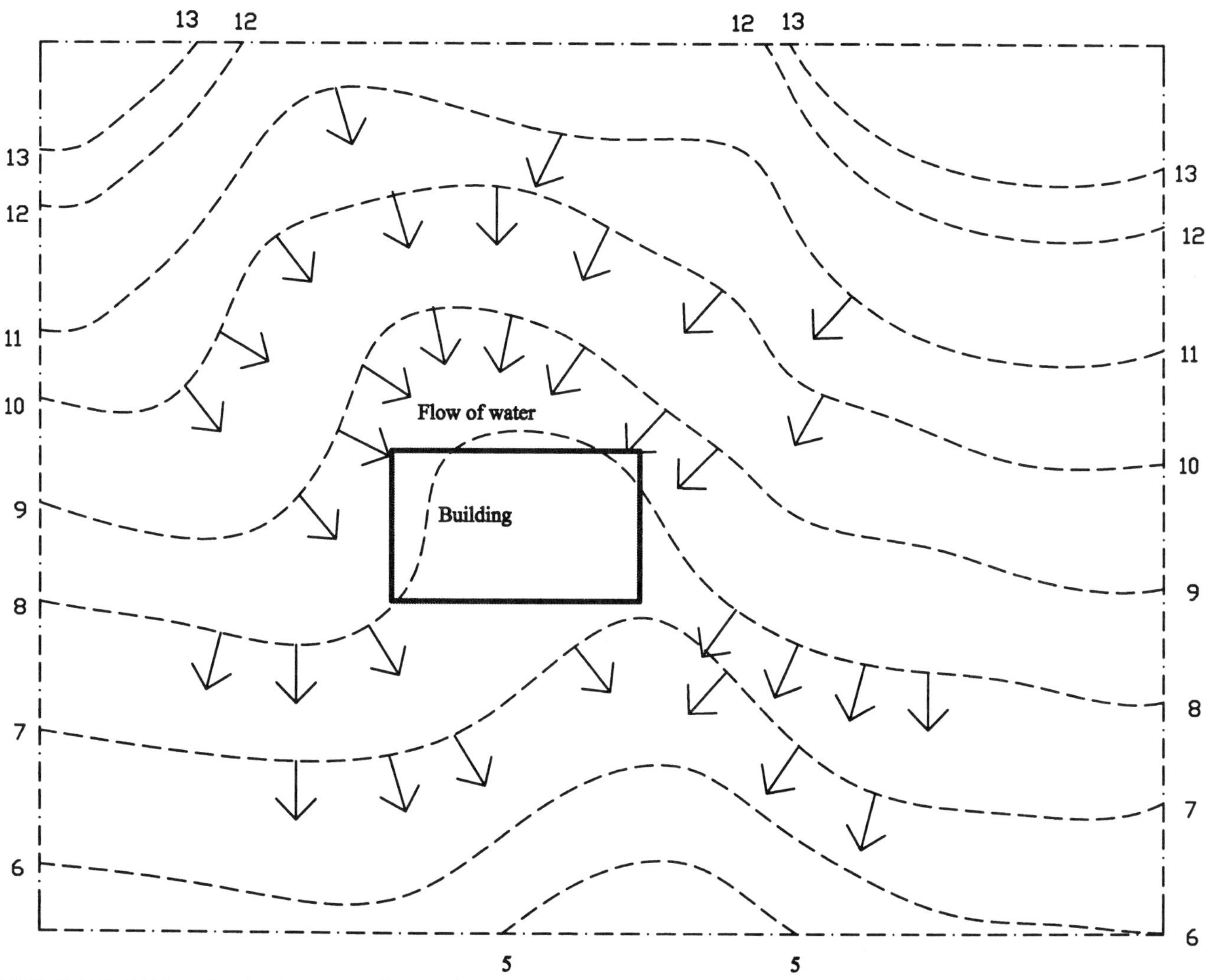

FIGURE 2.1(A) Original contour lines

- Drainage of large paved areas across pedestrian paths should be avoided. Catch basins and trench drains can be used to collect substantial quantities of runoff from parking lots and paved pedestrian areas.

SUBSURFACE RUNOFF

Subsurface runoff is the secondary method of removal of water from a site. This process allows the water to permeate the surface, be absorbed by the soil, and eventually become part of the groundwater supply and the aquifers. It is a very important contributor to the recharge of groundwater.

EVAPORATION AND TRANSPIRATION

A third method by which water is removed from a site is evaporation and transpiration. A site that receives plenty of sunlight will lose surface water through evaporation and reinforce the natural systems. Transpiration from vegetation at site is also helpful in removing surface water from site. Plants and vegetation absorb water during photosynthesis and transpire it to the atmosphere.

DELAY RUNOFF

The traditional method of storm water disposal in city areas aims at draining the water away as fast as possible, via gutters and storm sewers, to the natural

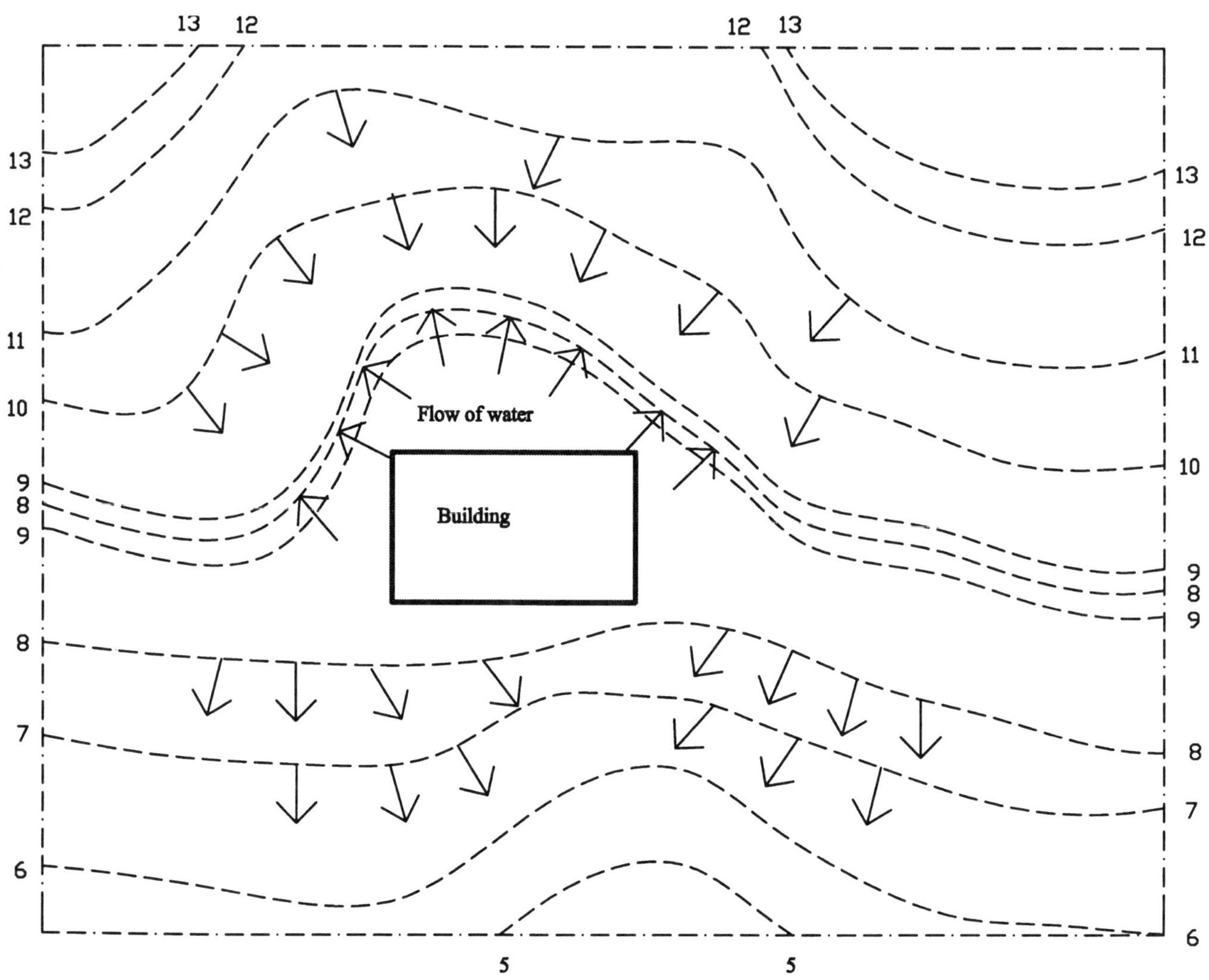

FIGURE 2.1(B) Revised contour lines

water bodies. This traditional approach, however, is reported to contribute to increased flooding and stream erosion. An alternative solution to this problem is local disposal of storm water at its source of runoff. Although this approach has been getting a lot of attention in recent years, it has been used in different parts of the world for a long time. This method, known as delayed runoff, can be used to control storm water from individual lots.

Advantages cited for delayed runoff include:

- recharge of groundwater;
- preservation and enhancement of natural vegetation;
- reduction of pollution to the natural water bodies;
- smaller storm sewers at low cost; and
- reduction of downstream flow peaks.

Disadvantages of the system include:

- eventual sealing of soil, leaving the property owners with a failed disposal system; and
- rise in groundwater level, causing flooding of basements or damage to building foundations.

2.2 SITE IRRIGATION

WATER FOR IRRIGATION

The amount of water required for landscaping depends on the variety and location of plants in a site. On an average, about 70 gallons of water per person

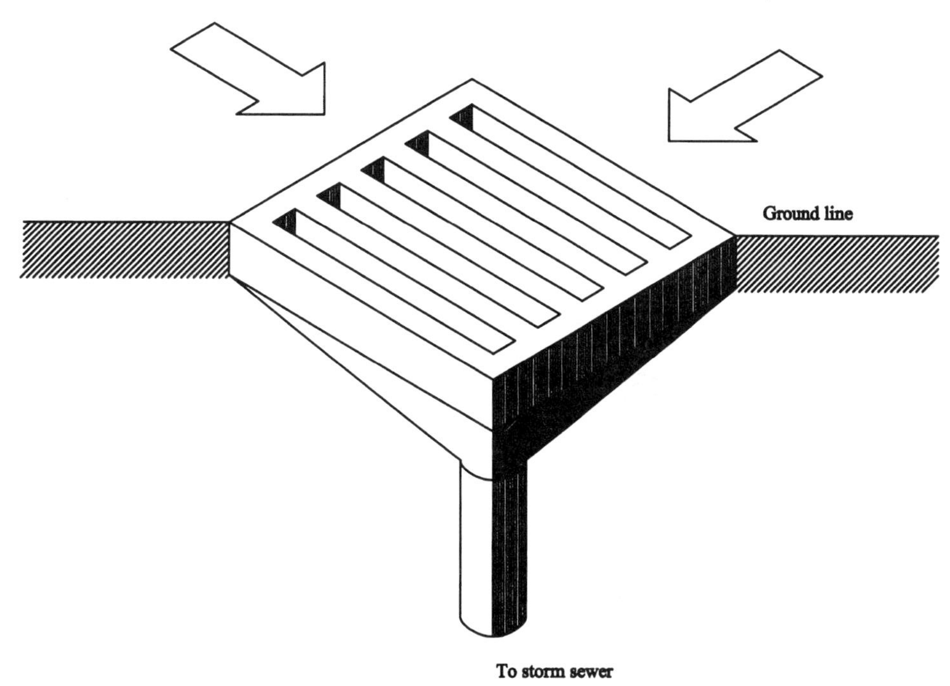

FIGURE 2.2(A) Area drain

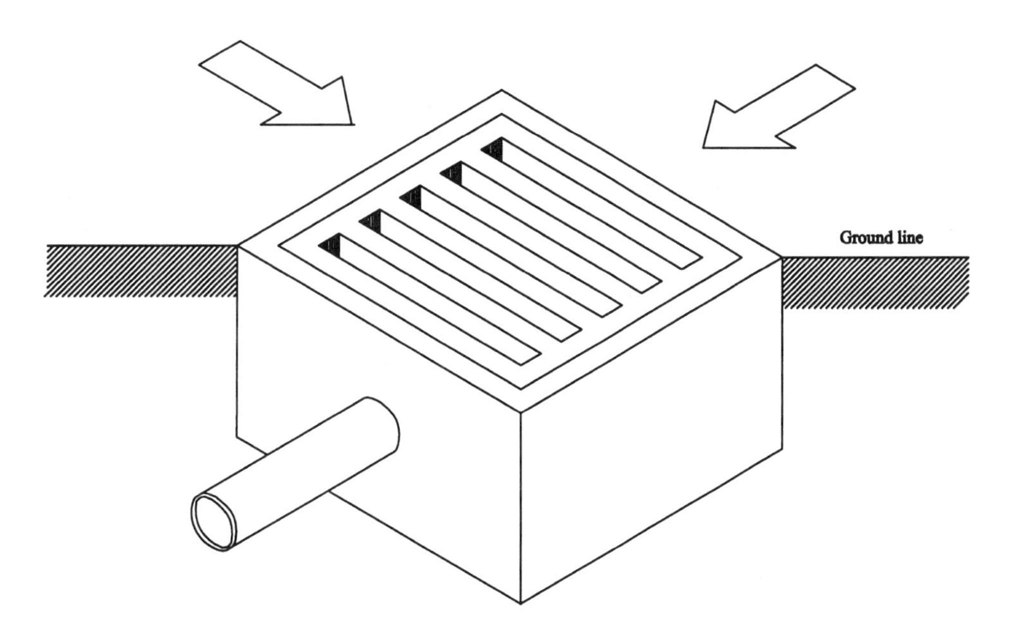

FIGURE 2.2(B) Catch basin

is used each day for outdoor use. Most of this water is utilized to maintain outdoor landscaping.

The quantity of water required for irrigation depends on type of soil and climatic conditions, as well as variety and location of plants. In order to optimize outdoor water use, it is a good idea to design a zone plan for the site based on the water requirements of the different types of trees, plants, and grass. Vegetation with similar water requirements may be grouped under a single irrigation zone. Such zoning will not only simplify the landscaping plan, but will also help to conserve water.

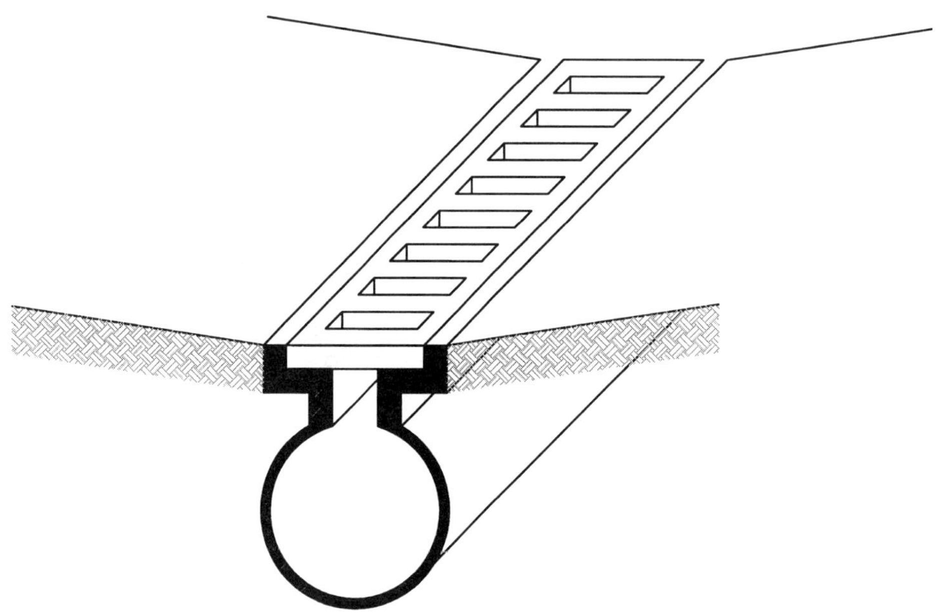

FIGURE 2.2(C) Trench drain

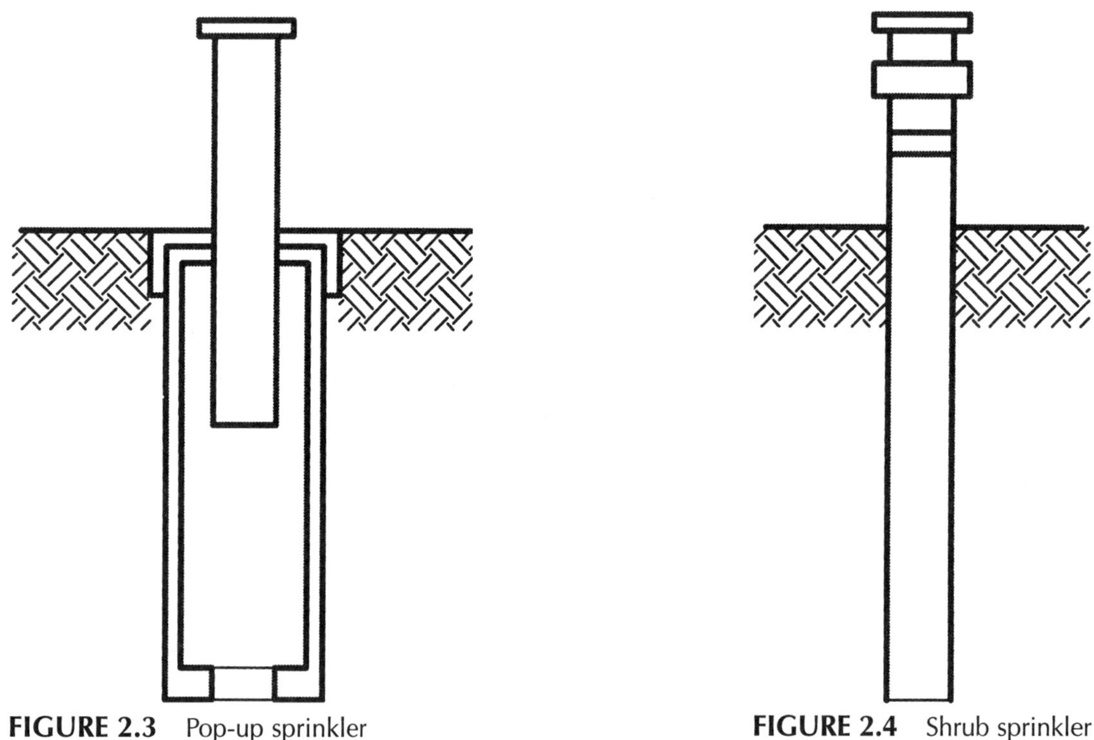

FIGURE 2.3 Pop-up sprinkler

FIGURE 2.4 Shrub sprinkler

SPRAY IRRIGATION

The most effective way to irrigate a lawn is by spray irrigation using automatic sprinklers. A spray irrigation system works very efficiently for a well-defined site with simple configuration. Narrow lawn areas with less than ten feet width are not recommended because of a possibility of overspray.

Types of Sprinklers Sprinklers for spray irrigation come in two basic types: pop-up and shrub. Pop-up sprinklers are installed below ground (Figure 2.3); part of the sprinkler rises above grade when the sprinkler is in operation and retracts back when it is shut off. Shrub sprinklers are installed above ground on top of a riser (Figure 2.4). It is advantageous to use pop-up sprinklers from the point of view of safety. It is easy to trip over, or fall on, a shrub sprinkler while taking care of the plants.

Height of a pop-up sprinkler when it rises above ground ranges from 3 to 12 inches. It includes a spring retraction for operating the sprinkler riser and a wiper seal around the riser stem to prevent leakage.

Types of Sprinkler Heads Sprinkler heads are divided into two types based on the method they use to distribute the water: fixed spray heads and rotors. Fixed spray heads spray a fan-shaped pattern of water, while rotors operate by rotating streams of water back and forth over the shrubs and plants.

Rotors usually cover more area than fixed spray heads, but cost more. If the area to be covered is more than eighteen feet, then it is more economical to use a fixed head spray. Rotors also require a higher water pressure to operate. They do not work properly at a pressure less than 40 psi.

Sprinkler Spacing The area covered by each sprinkler should overlap the area covered by the adjacent sprinkler to avoid creating dry spots. Maximum on-center distance between sprinklers should not exceed 1.2 times the radius of the spray head. For rotor type sprinklers, the spacing in feet should not exceed the operating pressure of the sprinklers in psi.

Backflow Preventer Local codes in most places require a backflow preventer to be installed on the irrigation system. This equipment helps to keep lawn fertilizers, weed killers, and other contaminants from being drawn into the potable water system. Manual or automatic valves cannot be substituted for backflow preventers.

2.3 SITE WATERSCAPE

POOLS AND FOUNTAINS

Water is often used for decorative or ornamental purposes in site design. Pools are usually constructed to evoke the impression of quiet water found in natural lakes; constructing fountains may simulate the essence of moving water such as waterfalls and geysers. The reflective quality of a pool adds tranquility to the environment and the moving quality of water imparted by a fountain enhances the visual interest of any landscape. Water fascinates all humans, regardless of their culture, ethnicity, and religious belief. It soothes and relaxes, inspires reflection, and is a source of pleasure.

Apart from aesthetic reasons, fountains also serve practical purposes. It may be effectively used for noise mitigation. The sounds of splashing, flowing, or moving water can mask adjacent irritating sources of noise in an urban environment. A fountain's ability to overcome noise is directly proportional to the sound produced by its water. A fountain may also serve as a source of evaporative cooling for an area with high ambient temperature and low humidity.

FOUNTAIN AND POOL EQUIPMENT

Mechanical systems of pools and fountains consist of three components: water effects system, water level control and drainage system, and filtration system. The water effects system of a fountain requires a pump, a discharge line to supply water, and a nozzle through which water will be discharged.

Nozzles There are four basic types of nozzles: aerating, spray head, smooth-bore, and formed. Aerating nozzles produce white, frothy water visible from a significant distance (Figure 2.5). This is accomplished by installing the nozzle under the water level and drawing a mixture of air and water through a perforated cover at the bottom of the nozzle. This type of nozzle requires a larger horsepower pump than other types of nozzles.

Clear and smooth water columns characterize a smooth-bore nozzle (Figure 2.6). The first part of the trajectory from such a nozzle is a solid-stream jet of water, which gradually breaks up into fine sprays as it reaches maximum height and descends downward.

A spray head nozzle produces delicate combinations of clear thin water jets (Figure 2.7). The distribution head of such a nozzle is usually circular or fan-shaped with multiple exposed tube nozzles or holes in the face.

A formed nozzle spurts out thin sheets of water (Figure 2.8). Size and shape of the water sheets vary according to the design of the nozzle. Unlike other nozzles, formed nozzles are not dependent on the

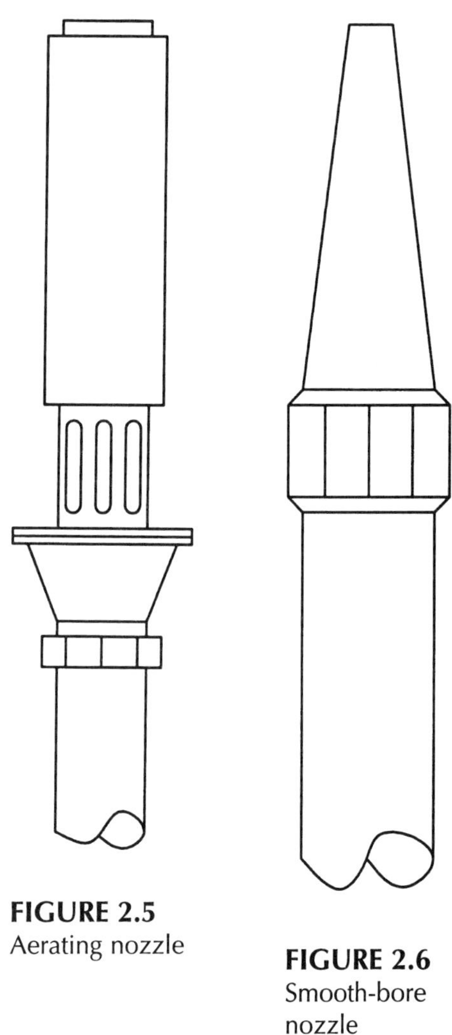

FIGURE 2.5 Aerating nozzle

FIGURE 2.6 Smooth-bore nozzle

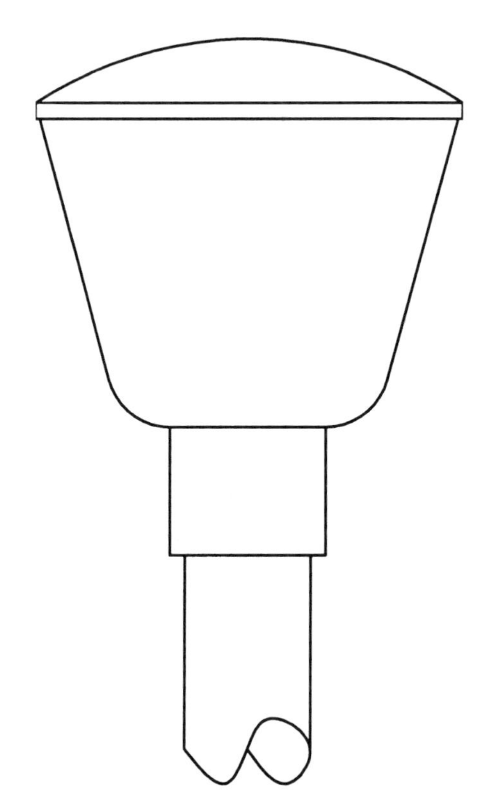

FIGURE 2.7 Spray head nozzle

level of water in pools; they can be located any distance above the water level.

Pumps Fountains nowadays are predominantly designed as closed systems; water held in basin is continually recirculated from basin to nozzle and back again to the basin. A pump mechanism is required to generate pressure to recirculate the water. Submersible pumps are used for small fountains requiring a discharge rate of 100 gpm or less; centrifugal pumps are used for fountains where large water volumes or high pressures are required.

FILTER SYSTEMS

Fountains are usually equipped with a filter system. It prevents damage to the nozzle and pump, and thereby reduces maintenance costs. A filter system

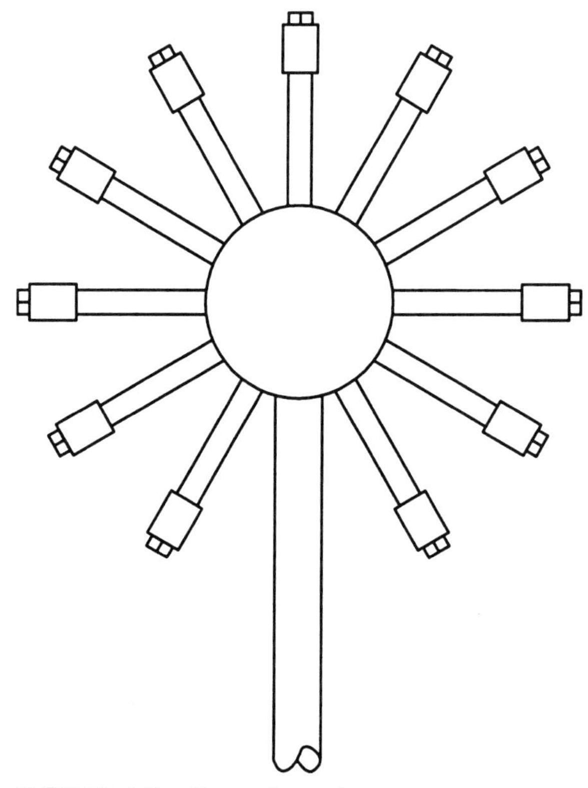

FIGURE 2.8 Formed nozzle

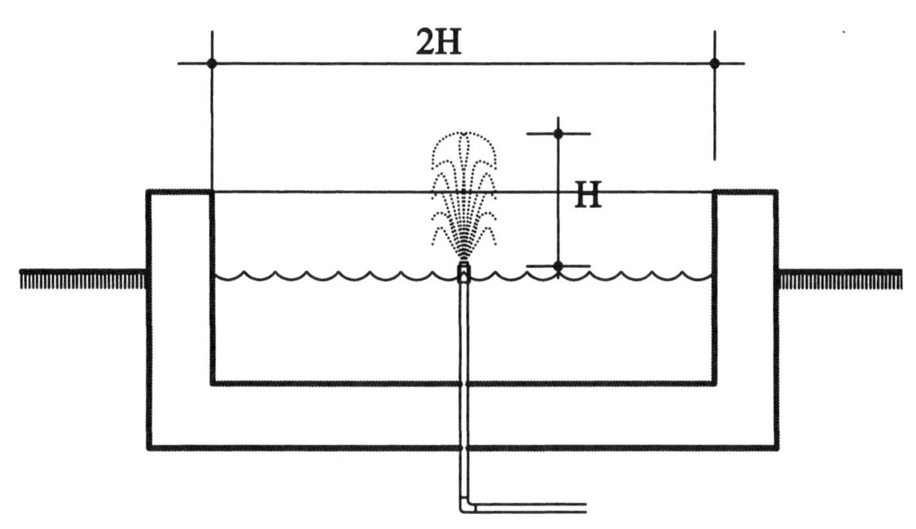

FIGURE 2.9 Sizing of fountain basins

consists of a pump and motor, a filtration medium, and a discharge line. A pump and motor for filter systems is a separate unit from the water effects system. These allow the filter system to operate independent of the water effects system. Cartridge filters, diatomaceous-earth filters, or permanent-media sand filters are most often used for filtration.

A filter system should be capable of recirculating the entire volume of pool water in 6 to 10 hours. If total volume of water of a basin containing 9,000 gallons is to be recirculated in 10 hours, the filter system must circulate 900 gallons per hour or 15 gallons per minute.

WATER LEVEL CONTROL

Water in a pool may be lost due to evaporation and splash, causing a fall in water level. This fall may cause damage to nozzles and pumps that require a minimum water protection for their continued operation. Water level controls are therefore required in fountains to replenish the lost water.

Appropriate water levels may be maintained using float controls or probe-type water level controls. A float control consists of a plastic float and a brass valve. The float, attached to an arm, operates the valve. The float is depressed when the water level falls, causing the valve to open and fresh water to flow into the fountain. This device is commonly used for small fountains.

Probe-type water level controls are used for large fountains. These devices consist of an electronic probe box for sensing water level, a relay control panel, and a solenoid-operated valve. The probe sends a signal to the relay panel when the water level drops below a minimum, which in turn sends a signal to the solenoid valve. This signal causes the valve to open, allowing fresh water to enter the fountain.

SIZING

Fountain Basin Fountain basins should be large enough to contain splashes from the nozzles and deep enough to ensure smooth and efficient operation of the equipment. Horizontal dimension of a fountain basin, with a single nozzle at the center, should not be less than twice the height of the water effect above the water surface (Figure 2.9). For windy locations, this dimension should be four times the height of the water effect.

Pump Size Water capacity of fountain pumps should be equal to the total capacity of all nozzles installed on the discharge line. A factor of safety to account for decrease in pump performance due to aging is usually added to the capacity of nozzles when pumps are specified. A common practice is to add 10 percent to the total nozzle capacity.

2.4 ROOF DRAINAGE

ROOF DRAINS

Sloping Roof Roof drains are required for disposal of rainwater collected on roofs of buildings. Rainwater from sloped roofs is first collected in gutters, which, in turn, direct the water to vertical leaders. The leader inlets should preferably be installed with leaf screens to prevent clogging of the system. If the vertical leader is not connected to a storm sewer, then it should be ensured that water is released at least three feet away from the foundation.

Rainwater may be drained without the use of gutters from a sloping roof if it has an overhang of three feet or more. A system without gutters requires that ground surface where drainage water strikes the ground be treated with gravel. The ground must also slope away from the structure on all sides at a minimum of 2 to 5 percent to carry water immediately away from the foundation.

TABLE 2.1

Sizing of Gutters

Slope of Gutter	*Diameter of Gutter (Inches)*	*Maximum Rainfall (Inches Per Hour)*				
		2	*3*	*4*	*5*	*6*
1/16-in	3	340	226	170	136	113
	4	720	480	360	288	240
	5	1,250	834	625	500	416
	6	1,920	—	960	768	640
	7	2,760	1,840	1,380	1,100	918
	8	3,980	2,655	1,990	1,590	1,325
	10	7,200	4,800	3,600	2,880	2,400
1/8-in.	3	480	320	240	192	160
	4	1,020	681	510	408	340
	5	1,760	1,172	880	704	587
	6	2,720	1,815	1,360	1,085	905
	7	3,900	2,600	1,950	1,560	1,300
	8	5,600	3,740	2,880	2,240	1,870
	10	10,200	6,800	5,100	4,080	3,400
1/4-in	3	680	454	340	272	226
	4	1,440	960	720	576	480
	5	2,500	1,668	1,250	1,000	434
	6	3,840	2,560	1,920	1,536	1,280
	7	5,520	3,680	2,760	2,205	1,840
	8	7,960	5,310	3,980	3,180	2,655
	10	14,400	9,600	7,200	5,750	4,800
1/2-in	3	960	640	480	384	320
	4	2,040	1,360	1,020	816	680
	5	3,540	2,360	1,770	1,415	1,180
	6	5,540	3,695	2,770	2,220	1,850
	7	7,800	5,200	3,900	3,120	2,600
	8	11,200	7,460	5,600	4,480	3,730
	10	20,000	13,330	10,000	8,000	6,660

Note: Figures in the field indicate horizontal projected roof areas in square feet.

TABLE 2.2

Sizing of Roof Drains and Leaders

Rainfall (Inches Per Hour)	Diameter of Drain or Leader (Inches)					
	2	3	4	5	6	8
1	2,880	8,800	18,400	34,600	54,000	116,000
2	1,440	4,400	9,200	17,300	27,000	58,000
3	960	2,930	6,130	11,530	17,995	38,660
4	720	2,200	4,600	8,650	13,500	29,000
5	575	1,760	3,680	6,920	10,800	23,200
6	480	1,470	3,070	5,765	9,000	19,315
7	410	1,260	2,630	4,945	7,715	16,570
8	360	1,100	2,300	4,325	6,750	14,500
9	320	980	2,045	3,845	6,000	12,890
10	290	880	1,840	3,460	5,400	11,600
11	260	800	1,675	3,145	4,910	10,545
12	240	730	1,530	2,880	4,500	9,660

Note: Figures in the field indicate horizontal projected roof areas in square feet.

Flat Roof For collection of rainwater from a flat roof, a gentle slope should be provided on the built-up roof so that the water can be carried to a vertical roof drain. Tables 2.1, 2.2, and 2.3 can be used for sizing gutters, drains, leaders, and horizontal storm-water pipes, respectively.

Storm sewers are typically used for carrying the rainwater discharged from the roof. These pipes normally have multiple storm-water collection points (catch basins or area basins). They are interconnected through a series of pipes that carry water to a main collector line. The main collector transports the water to a natural creek or other water bodies, if permitted by local code. Otherwise, a dry-well system has to be designed and constructed in accordance with local codes. Table 2.3 can also be used for sizing storm sewers.

Roofs in Cold Climates Ice buildup on a sloped roof is a major problem in cold climatic conditions. This ice ridge on the edge of a roof is known as ice dam. The formation of an ice dam is the result of a complex interaction between amount of heat loss from a house, snow cover, and outside air temperature. Snow will melt on the part of the roof through which there is a heat loss. As the melted snow reaches the portion of the roof that is below 32°F it freezes and forms an ice dam (Figure 2.10). It grows in size as it is fed by more and more melting snow from above. Melted snow will back up behind the ice dam and remain as a liquid. This stagnant water will find cracks and openings on the exterior roof surface and will, eventually, find its way inside the building.

In order to prevent the buildup of ice dams, the ceiling/roof insulation has to be increased to cut down on heat loss by conduction. Some state codes require insulation with an R-value of 38 above the ceiling. The ceiling also has to be made airtight so that no warm air can flow from inside the house to the attic space.

Immediate actions for preventing the formation of ice dams include removal of snow from the roof using a roof rake. If an ice dam has already formed and it cannot be removed immediately, the least that can be done is to make channels through the ice dam so that the trapped water drains off. Hosing the dam with tap water will make the channels.

Drainage Below Grade A sump is required to be installed in a building with a subsurface storm drain in the basement. Storm water received by the sump is pumped out, to be eventually discharged either to the storm sewer or to a drywell.

TABLE 2.3

Sizing of Horizontal Storm-Water Pipes

Slope of Pipe	*Diameter of Pipe (Inches)*	*Maximum Rainfall (Inches Per Hour)*				
		2	*3*	*4*	*5*	*6*
1/8-in	3	1,644	1,096	822	647	548
	4	3,760	2,506	1,880	1,504	1,253
	5	6,680	4,453	3,340	2,672	2,227
	6	10,700	7,133	5,350	4,280	3,566
	8	23,000	15,330	11,500	9,200	7,600
	10	41,400	27,600	20,700	16,580	13,800
	12	66,600	44,400	33,300	26,650	22,200
	15	109,000	72,800	59,500	47,600	39,650
1/4-in	3	2,320	1,546	1,160	928	773
	4	5,300	3,533	2,650	2,120	1,766
	5	9,440	6,293	4,720	3,776	3,146
	6	15,100	10,066	7,550	6,040	5,033
	8	32,600	21,733	16,300	13,040	10,866
	10	58,400	38,950	29,200	23,350	19,450
	12	94,000	62,600	47,000	37,600	31,350
	15	168,000	112,000	84,000	67,250	56,000
1/2-in.	3	3,288	2,295	1,644	1,310	1,096
	4	7,520	5,010	3,760	3,010	2,500
	5	13,360	8,900	6,680	5,320	4,450
	6	21,400	13,700	10,700	8,580	7,140
	8	46,000	30,650	23,000	18,400	15,320
	10	82,800	55,200	41,400	33,150	27,600
	12	133,200	88,800	66,600	53,200	44,400
	15	238,000	158,800	119,000	95,300	79,250

Note: Figures in the field indicates horizontal projected roof areas in square feet.

2.5 PARKING LOT

DESIGN

One of the major components of building design and site improvement is to make provision for parking vehicles. In order to organize parking efficiently and to provide adequate spaces, one must be aware of the dimensions of the vehicles.

Parking lot sizes are measured in numbers of vehicles to be accommodated. Lots that accommodate 100 to 200 cars are efficient and practical. For efficient land use, a self-parking lot should provide about 300 square feet for each vehicle space. The size and capacity of a parking lot, however, should be tailored to the actual requirements.

Maximum capacity of a parking lot can be achieved by developing parking modules running parallel to the longer dimension of the site. A parking module includes a drive path with parking on either side. Cars may be parked at angles ranging from 0 to 90 degrees (Figures 2.11–2.14). Width of a parking module becomes progressively smaller as the angle of parking decreases. Width of parking stall sizes range from a minimum of 8′-6″ to a maximum of 12′-0″. Typical parking lot capacity figures at different angles of parking are given in Table 2.4.

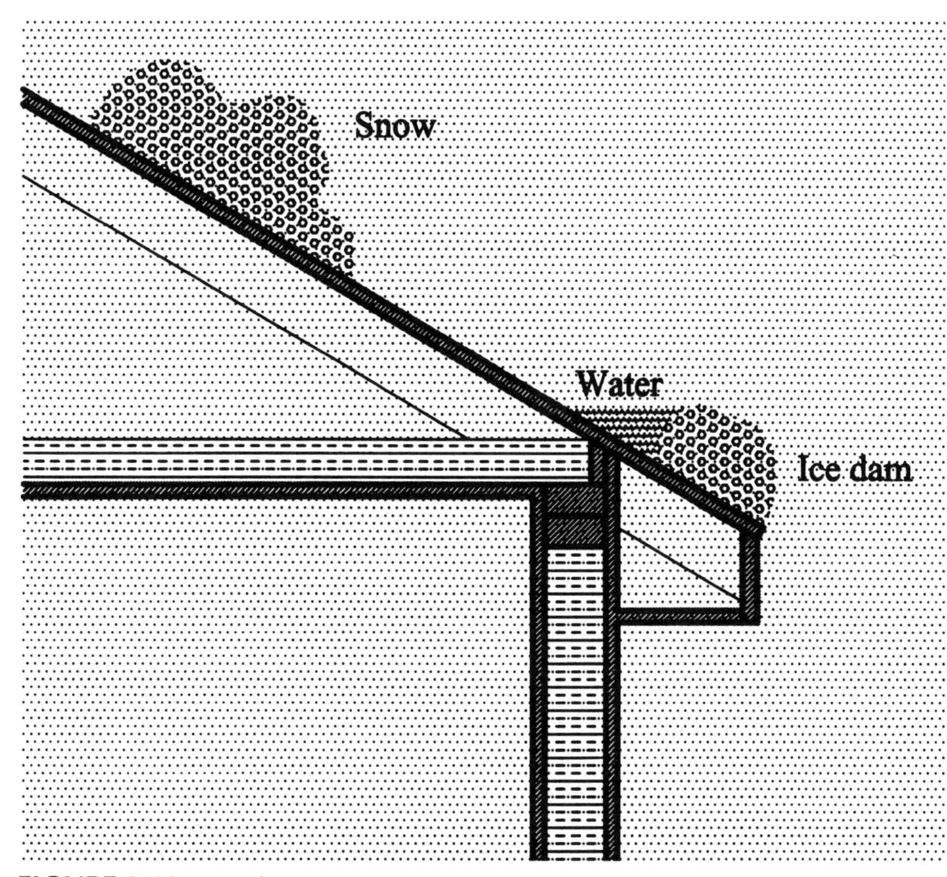

FIGURE 2.10 Ice dam

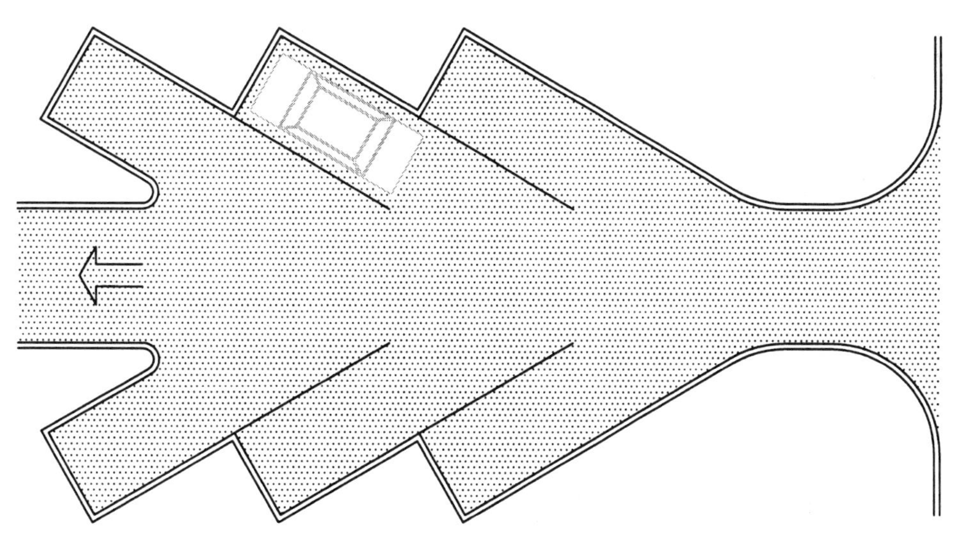

FIGURE 2.11 Parking at 30° angle

DRAINAGE

Adequate drainage of parking lots is important for preventing damage or inconvenience to abutting property and/or public streets and alleys. They are usually required to be improved with a compacted gravel base surfaced with an all-weather material. Maximum desirable grade of a parking lot in any direction should not exceed 5 percent. Steeper slopes make it difficult to maneuver cars in slippery

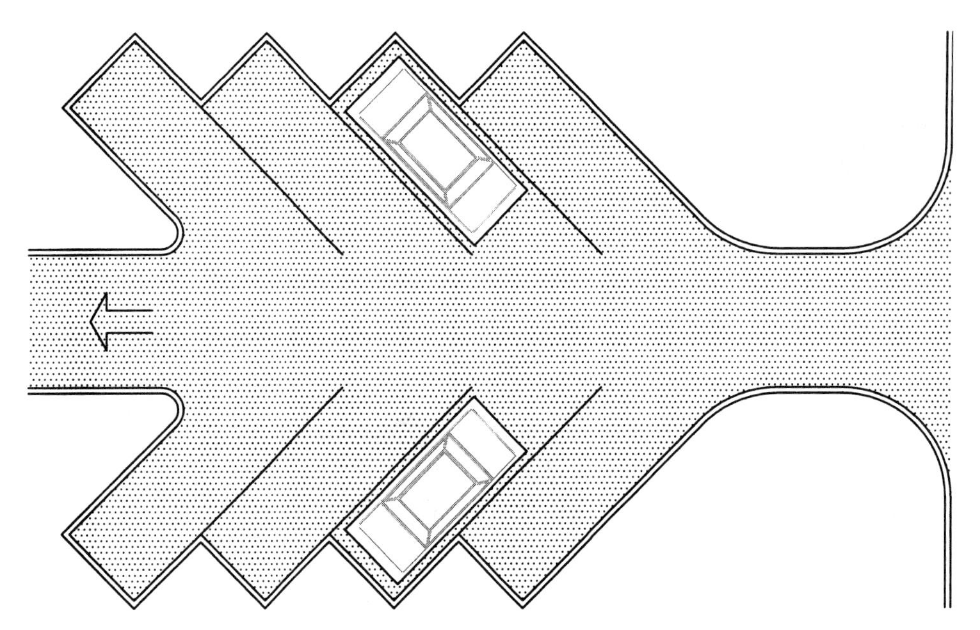

FIGURE 2.12 Parking at 45° angle

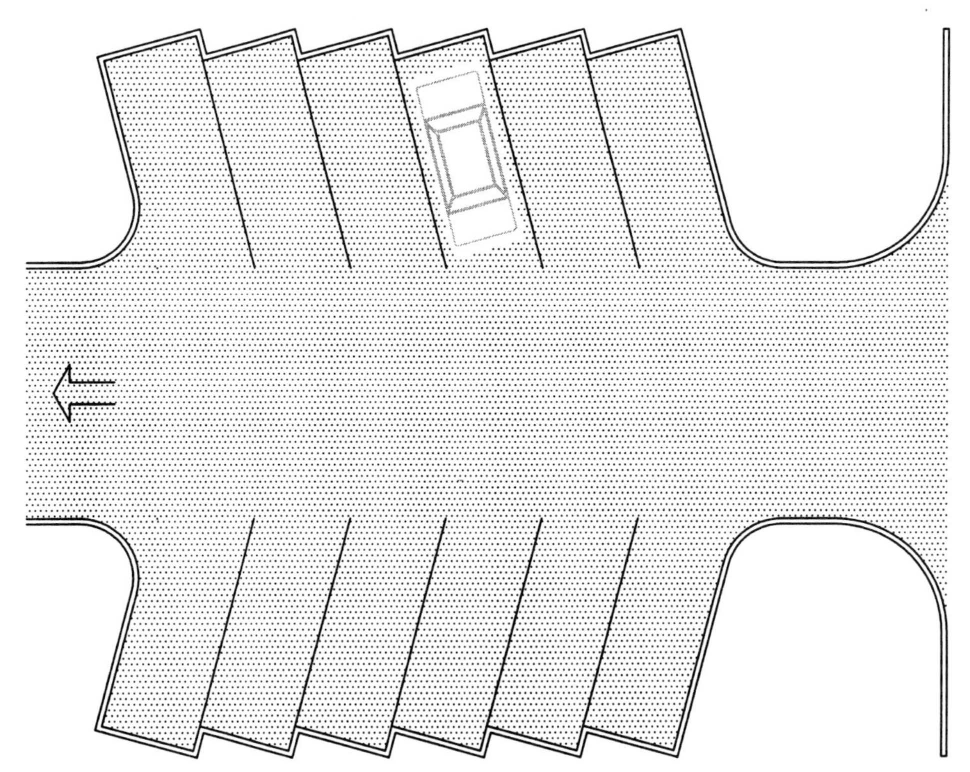

FIGURE 2.13 Parking at 75° angle

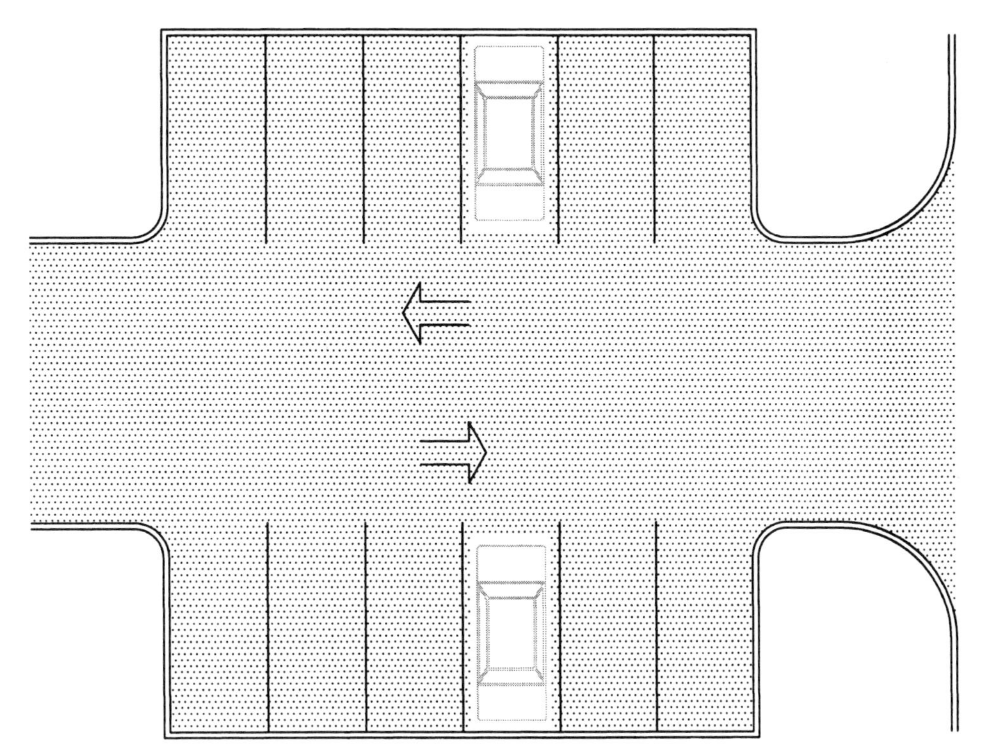

FIGURE 2.14 Parking at 90° angle

TABLE 2.4

Maximum Capacity of Parking Lots at Different Parking Angles

Parking Angle (Degrees)	*Curb Length Per Car (Feet)*	*Depth of Stall (Feet)*	*Width of Aisle (Feet)*	*Gross Area Per Car (Square Feet)*	*Unit Parking Depth (Feet)*	*Approx. Number of Cars Per Acre*
0	22	8	12	308	28	141
30	17	16.4	12	380.8	44.8	114
45	12	18.7	12	296.4	49.4	147
60	9.8	19.8	14.5	265.1	54.1	164
75	8.8	19.6	23	273.7	62.2	159
90	8.5	18	24	255	60	171

weather. Slope, however, should not be less than 1 percent in all directions to develop a positive drainage pattern.

Some local codes require the use of storm-water retention systems for large parking lots. It helps to manage storm-water runoff and prevent downstream flooding. Storm water can be retained in catch basins and eventually directed to the storm sewers.

REVIEW QUESTIONS

1. What are the basic objectives of a subsurface drainage system?
2. What is delay runoff?
3. Maximum recommended on-center distance between irrigation sprinklers is _____ .
4. What is an ice dam?
5. Rainwater from a vertical leader should be released at least _____ feet away from the foundation of a building.

ANSWERS

1. Collection, transfer, and disposal of surface water at site.
2. It is the disposal of surface water at its source of runoff.
3. 1.2 times the radius of the spray head of a sprinkler.
4. It is an ice ridge on the edge of the roof of a building.
5. 3.

Building Plumbing

A mountain village lost in snow . . . under the drifts of a sound of water.

Shiki

This chapter deals with water demand, distribution, and drainage. Piping is needed for both water supply and drainage. The chapter elaborates on the different types of materials for each purpose and the fittings, fixtures, and other accessories required for efficient installation and functioning of the systems. Issues related to the sizing of supply and drainage pipes are discussed at length. The chapter ends with some thoughts on conservation and efficient use of water.

3.0 WATER DEMAND

The first step in the selection of a suitable water supply source is deciding the demand that will be placed on it. Water demand refers to the quantity of potable water required by the inhabitants of a building. There are two different types of demand: average daily demand and peak demand.

Average daily water demand is the total quantity of water required by a building during the period when it is occupied. It depends on the functions served by the building, the types of fixtures used, and the total hours of operation of a facility. Table 3.1 presents a summary of average daily water demand by different types of establishments. An average pattern of home end uses in the United States range from about 4 gppd for drinking and cooking to 30 gppd for flushing toilets (refer to Table 1.2 in Chapter 1).

Peak demand refers to the rate of water use during critical periods of the day when water consumption is the highest in a building. This is also known as maximum momentary demand, or maxmo. This rate must be determined for the process of sizing water-supply piping. The total quantity of water used by a household may be distributed over only a few hours of the day, during which period actual water use is higher than the mean rate shown in Table 3.1.

HOT WATER DEMAND

End uses related to bathing, washing, laundry, and dishwashing require hot water. The average usage of hot water in the United States is assumed to be 20 gppd for a family of two persons. An additional 15 gppd will have to be added for an increase in the number of family members. Hourly demand for domestic hot water is considered to be 2.1 gallons per

TABLE 3.1

Average Daily Water Demand

Type of Facility	*Water Demand (In Gallons Per Day)*
Airport (per passenger)	3–5
Camps:	
Construction (per worker)	50
Day without meal (per camper)	15
Luxury (per camper)	100–150
Resorts, day and night, with limited plumbing (per camper)	50
Tourist with central bath and toilet facilities (per person)	35
Cottages with seasonal occupancy (per resident)	50
Country club (per resident member)	100
Country club (per nonresident member present)	25
Courts, tourist, with individual bath units (per person)	50
Dwellings:	
Boardinghouses (per boarder)	50
Luxury (per person)	100–150
Multifamily (per resident)	60
Rooming houses (per resident)	60
Single-family, indoor use (per resident)	70
Single-family, outdoor use (per resident)	70
Estates (per resident)	100–150
Factories (per person per shift)	15–35

TABLE 3.1 (Continued)

Average Daily Water Demand

Type of Facility	*Water Demand (In Gallons Per Day)*
Highway rest areas (per person)	5
Hotels with private baths (two persons per room)	60
Hotels without baths (per person)	50
Institutions (per person)	75–125
Hospitals (per bed)	250–400
Laundries, self service (per customer)	50
Livestock (per animal):	
Cattle (drinking)	12
Dairy (drinking and servicing)	35
Goat (drinking)	2
Hog (drinking)	4
Horse (drinking)	12
Mule (drinking)	12
Sheep (drinking)	2
Steer (drinking)	12
Motels with bath, toilet, and kitchen facilities (per bed space)	50
Motels with bath and toilet (per bed space)	40
Parks:	
Overnight with flush toilets (per camper)	25
Trailers with individual baths, no sewer connection (per trailer)	25
Trailers with individual baths, connected to sewer (per person)	50
Picnic facilities:	
With bathhouses, showers, and flush toilets (per person)	20
With toilet facilities only (per person)	10
Poultry:	
Chickens (per 100)	5–10
Turkeys (per 100)	10–18
Public baths (per bather)	10
Restaurants with toilet facilities (per patron)	7–10
Restaurants without toilet facilities (per patron)	2.5–3
Restaurants with bars and cocktail lounge (additional quantity per patron)	2
Schools:	
Boarding (per student)	75–100
Day, with cafeteria, gymnasiums, and showers (per student)	25
Day, with cafeteria, but no gymnasiums or showers (per student)	20
Day, without cafeteria, gymnasiums, or showers (per student)	15
Service stations (per vehicle)	10
Stores (per rest room)	400
Swimming pools (per swimmer)	10
Theaters, movie (per seat)	5
Workers, day (per person per shift)	15

person. Table 3.2 presents an average usage rate of residential hot water.

Hot water is generated using a wide variety of devices and different types of energy sources. Water is defined as hot when it attains a temperature of 110°F or higher. Energy sources used for generating hot water include electricity, natural gas, oil, and solar energy. Water heaters that use electricity may either be resistance type or heat pump. Heat pump water heaters use electricity to move rather than create heat. The heat source is the outside air or the air where the unit is located. Refrigerants and compressors transfer heat into an insulated storage tank.

TABLE 3.2

Residential Hot Water Consumption

Usage	*Consumption (in gppd)*
Bath	25
Shower (5 minutes)	6
Wash (using lavatory)	3
Dishwasher	10
Clothes washer	36
Kitchen sink	6

Water heaters may either be storage type or instantaneous. Storage type water heaters consist of an insulated storage tank and a device for heating. The heating device heats up a fixed quantity of water that is stored in the tank and used when required. An instantaneous water heater converts cold water instantly to hot water as it passes through it. The process continues as long as a fixture draws the hot water. It eliminates standby losses that occur in storage tanks and piping.

Water heaters are rated by their tank capacity in gallons and recovery rate in gallons per hour. The recovery rate is simply the number of gallons per hour that a water heater can produce. This rate is usually based on a cold water inlet temperature of about 50°F and a final temperature ranging from 110°F to 140°F. For a given load, recovery rate of a water heater is inversely proportional to the hot water storage tank size.

The usable quantity of hot water available from a storage tank is usually assumed to be 70 percent, unless mentioned otherwise. This means that if the actual capacity of a water heater tank is 100 gallons, the quantity available at a desired temperature would be about 70 gallons. Table 3.3 presents a guideline for sizing residential water heater tanks, considering both household size and the hot water plumbing fixtures.

TABLE 3.3

Sizing of Domestic Water Heater Tank

Hot Water Load	*Number*
Size of household	Indicate total number persons in the family
Bathtubs or showers	Indicate total number of fixtures
Washing machines	Indicate total number of fixtures
Automatic dishwashers	Indicate total number of fixtures
Total load	**Sum of the above numbers**

Total Load	*Capacity of Gas Water Heater Tank (In Gallons)*	*Total Load*	*Capacity of Electric Water Heater Tank (In Gallons)*
3 or less	30	4 or less	40
4 or 5	40	5 to 7	60
6	50	8 or more	80
7 or more	60		

Water heating typically represents about 15 to 25 per cent of the total household energy use in the United States. It is, therefore, important to ensure that the heater is operated using the most efficient energy source. Annual quantity of fuel required for operating a water heater can be calculated using the following formula:

$$F = Q/(V \cdot n) \qquad (1)$$

where F = total quantity of fuel required per year, Q = total quantity of heat required to produce hot water per year in British thermal units (BTU), V = heat content per unit (gallon, therm, or kWh) of the fuel, and n = percent efficiency of the water heater.

Unit of measurement of heat is BTU. Specific heat of water (i.e., heat required to raise the temperature of one pound of water 1°F) is 1 BTU. Therefore, total quantity of heat required per year to produce hot water can be calculated using the following formula:

$$Q = G \cdot 8.33 \cdot \text{TD} \qquad (2)$$

where Q = BTU per year, G = gallons of hot water produced per year, and TD = temperature difference between cold supply water and hot water.

Total quantity of fuel required per year to produce hot water is calculated using the following formula:

$$E = Q/(F \cdot e) \qquad (3)$$

where E = quantity of fuel (in gallons, therm, or kWh), F = fuel heat content in BTU, and e = efficiency of the heater. Fuel heat content of different types of energy sources is shown in Table 3.4.

Once the total quantity of fuel required per year to produce hot water has been established, the total cost of operation can be calculated using the unit cost of the fuel type. The annual cost of production of hot water can be calculated using different types of fuel. This will form a simple basis for choosing the type of water heater to be installed according to fuel type.

The most economical choice of fuel is usually natural gas. Natural gas costs about two-thirds less per million BTUs than electricity. For locations where natural gas is not available, electricity or oil is commonly used for water heating. Initial cost of heat pump water heaters is high, but the efficiency of such a unit is much higher than an electrical resistance water heater. Solar water heaters use a renewable resource. This system is the most economical way to produce hot water during the period when it is abundantly available. Solar energy can meet most of the summertime demand for domestic hot water throughout the United States. The system needs to be protected against freezing in winter months, especially in the northern part of the country. Solar water heaters may be either active or passive. The active system relies on a pump or controller to move the water or collector fluid from the roof collector to a storage tank located in the house; in the passive system the storage tank rests directly upon the roof.

3.1 WATER DISTRIBUTION

MUNICIPAL STREET MAIN

Potable water is distributed in urban areas through municipal street mains. These are large pipes that usually run underground below the streets. Supplied by the water companies, the water flows under pressure that normally ranges between 50 to 70 pounds per square inch (psi) by the time it gets to a building.

TABLE 3.4

Fuel Heat Content

Fuel	*Unit*	*Heat Content (In BTU)*
Coal	Pound	14,000
Electricity	kWh	3,400
Natural gas	Therm (100 cu. ft.)	100,000
Oil #2	Gallon	140,000
Propane	Gallon	95,500

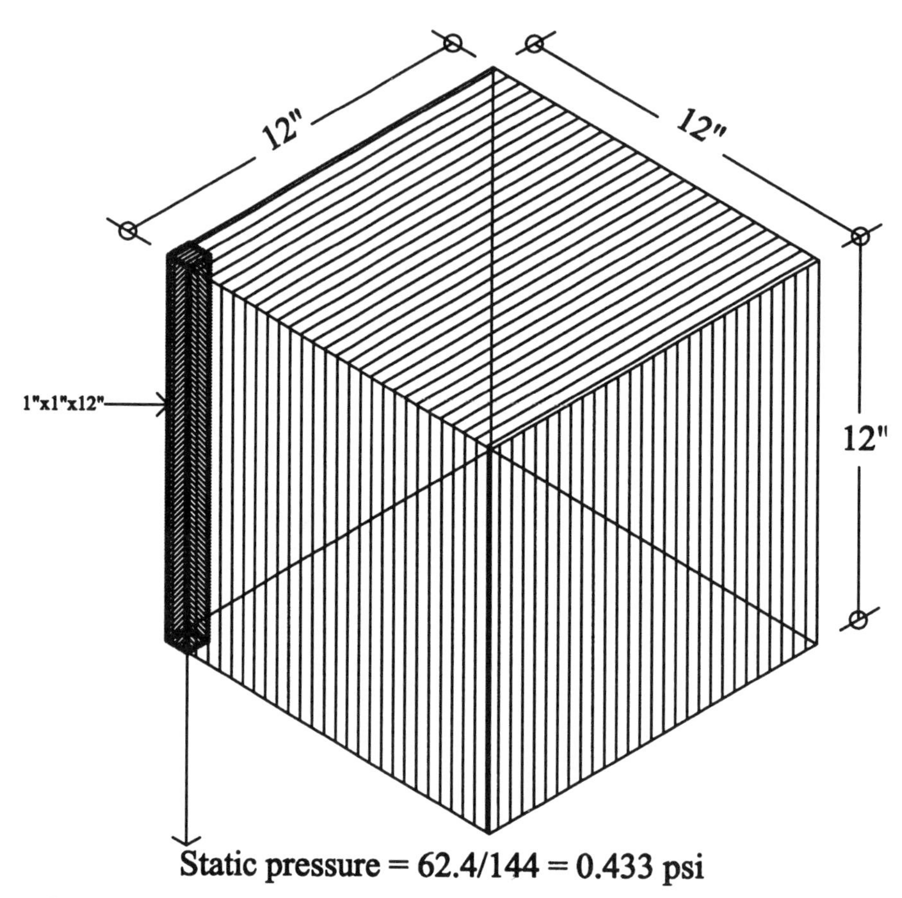

FIGURE 3.1 Static pressure

The supply main pipe of each building must be buried below frost line in cold climates so that it does not freeze. This depth can vary from about 2 to 7 feet depending on the climate.

The pressure of water supply to a building must be great enough to overcome the frictional resistance offered by the distribution system and the static pressure (static head) of water. Moreover, there must be adequate pressure left for the plumbing fixtures to operate properly. The water pressures in the municipal street main are usually adequate for supply to the highest and remotest fixtures in low-rise buildings.

Static Pressure Static pressure (static head) is pressure exerted by water standing in vertical piping. The weight of a cubic foot of water is 62.4 pounds. Therefore, downward pressure exerted by a 1-square-inch column of water, 1 foot in height, is 0.433 psi (Figure 3.1). If the supply pressure in a municipal street main is 50 psi, it can withstand the pressure exerted by a column of water more than 115 feet in height. Considering the pressure necessary to operate the fixtures on the top floor of a building, a 50-psi pressure is powerful enough to supply water to fixtures more than 50 feet above the supply main level.

Upfeed Distribution When water is fed to fixtures in a building by the incoming pressure of the water, it is called upfeed distribution. As stated earlier, this method is good for buildings up to five or six stories high. For upfeed distribution to work in taller buildings, additional pumps have to be installed to increase pressure. The augmented system is known as pumped upfeed distribution. This method is good for medium-size buildings that cannot rely on street main pressure.

Upfeed distribution for medium-size buildings can also be provided using hydro-pneumatic tanks. In this system, water is forced into hermetically sealed

TABLE 3.5

Comparison of Different Pipe Materials for Water Supply

Material	*Major Advantages*	*Major Disadvantages*
Copper	Long lasting Easy to put together and dismantle Resists attacks by most acids Thin-walled Lightweighted Low frictional resistance	Very expensive Requires soldering
Galvanized steel	Strong Relatively inexpensive Resistant to rough handling High pressure rating	Heavy Susceptible to corrosion High frictional resistance
Plastic	Inexpensive Lightweight Easy to install Very low frictional resistance Corrosion resistant	High thermal expansion Low strength Brittle when cold Easily scratched

vessels, compressing the air within. This captive, compressed air maintains the required pressure and forces the water to flow into the distribution network.

Downfeed Distribution Downfeed distribution systems may be designed for buildings more than six stories in height. In this system, water from a street main or suction tank is pumped to the roof of the building to storage tanks. The water from the storage tanks serves the floors below due to the force of gravity.

The water pressure required to serve the top floor of a building is usually 25 psi. Maximum pressure exerted on the fixtures of the bottom floor is recommended not to exceed 50 psi. A water pressure above 80 psi can damage the fixture valves. In order to keep the water pressure within proper limits, buildings higher than 15 stories in height are divided into more than one zone. Each zone should have a reservoir placed approximately 35 feet above the floor of the top level to produce a minimum pressure required to operate the fixtures. Use of pressure-reducing valves on the water main at bottom-floor level of each zone is recommended in order to avoid exceeding the maximum pressure limit.

SUPPLY PIPING MATERIALS

Water pipes and fittings may be of brass, black steel, copper, galvanized steel, or plastic (see Table 3.5). However, the local plumbing code may specify the type of materials that may be used for each particular piping system.

Steel and Galvanized Steel Steel pipe may be used for supply when water is noncorrosive. It is made from mild carbon steel as either welded or seamless pipe. Unprotected steel pipe rusts upon exposure to the atmosphere and moisture. In order to provide a protective coating, the steel pipe is dipped in a hot bath of molten zinc. This process is known as galvanizing. Nominal sizes of galvanized steel pipe range from 1/8 inch to 12 inch, in several wall thicknesses. The pipe wall thickness is usually described using the terms Schedule 40, for standard wall, and Schedule 80, for extra strong wall. Schedule 40 is normally used for typical plumbing applications.

Steel and galvanized steel pipes are connected with threaded fittings. For this reason, the pipes are normally manufactured with threads on both ends and supplied with a coupling threaded onto one end.

Steel pipes in sizes larger than 4 inches are usually welded or connected by bolted flanges.

Copper Copper pipe is substantially resistant to the chemical attack of many acids, salts, and bases. It is a durable pipe that handles high water pressure loads and is relatively easy to work with. However, copper is expensive and can cost much more than either steel or plastic pipe.

The pipes are manufactured from pure copper by drawing a heated billet through a die. There are four different types of copper pipe or tubing available according to wall thickness: K, L, M, and DWV. Type K tubing has the heaviest wall and type DWV has the thinnest wall. Types K, L, and M are used for water supply while DWV is used for drainage. K and L types are available both as rigid and soft copper tubes, while M and DWV types are available only as rigid copper tubes. They are marked with color codes: green for type K, blue for type L, red for type M, and yellow for type DWV.

Copper pipes fit together with lead-free, solid-core solder using solder type fittings. The soldering process involves heating the surfaces and then applying molten solder (usually a tin-antimony alloy) between the surfaces of the pipe and the fitting. Solder joints depend on capillary action drawing free-flowing molten solder into the narrow clearance between the fitting and the pipe. Molten solder metal is drawn into the joint by capillary action regardless of whether the solder flow is upward, downward, or horizontal. Solder type fittings are used above ground for both rigid and soft copper tubes. Shapes and patterns of these fittings are similar to those used for galvanized pipes.

Mechanical joints involving flared tube ends are frequently used for soft copper tubes or when the tubing is subject to vibration. The joint is made by slipping a flare-type nut over the end of the copper tube, flaring the end of the tube, and then screwing the flare nut onto a threaded fitting. A compression fitting may also be used to join copper tubes.

Plastic Plastic pipes are available in different varieties. They are produced from synthetic resins derived from fossil fuels such as coal and petroleum. There are seven types of plastic: (1) polyvinyl chloride (PVC), (2) chlorinated polyvinyl chloride (CPVC), (3) acrylonitrile butadiene styrene (ABS), (4) polyethylene (PE), (5) styrene rubber (SR), (6) polybutylene (PB), and (7) polypropylene (PP). Out of these seven types, four are commonly used for plumbing pipes and fittings: (1) acrylonitrile butadiene styrene (ABS), (2) polyvinyl chloride (PVC), (3) chlorinated polyvinyl chloride (CPVC), and (4) polyethylene (PE). ABS is commonly used for drainage piping; PVC is used to distribute cold water; CPVC is used for distribution of hot and cold water; and PE is used for cold water supply piping, below ground, from a municipal water main to a building.

Plastic piping is lightweight, inexpensive, and resistant to corrosion and weathering. Smooth interior walls of plastic piping offer low frictional resistance to water supply. However, it has a low resistance to heat, a high coefficient of expansion, and low crush resistance. Pressure ratings of a plastic pipe are also lower than metal pipes. The plastic piping that can withstand higher temperature is CPVC. It is a rigid, high-strength thermoplastic polymer (polyvinyl dichloride) that is practically inert toward water, inorganic reagents, hydrocarbons, and alcohol over a broad temperature range. The CPVC pipe is suitable for carrying hot water limited to temperatures not greater than 180°F.

The fittings for plastic pipes are similar to those used for galvanized steel pipes. Pipes are joined with the fitting using a process called solvent welding. The material used for the purpose is known as solvent cement. It softens the piping material on the outside and the fitting material on the inside. When joined together under proper conditions, the surfaces run together and fuse, producing a joint that is as strong as the pipe itself.

Joints in PE pipes are made with tapered and notched inserts of either polystyrene or galvanized steel. After the insert has been slipped into the ends, the pipe is compressed against it by means of metal steel clamps.

Threaded joints are also available for plastic pipes. They are usually used for Schedule 40 PVC pipes.

WATER SUPPLY ACCESSORIES AND CONTROLS

Valve It is a fitting used on a piping system to control the flow of fluid within that system in one or more ways. It is desirable to install a valve to control

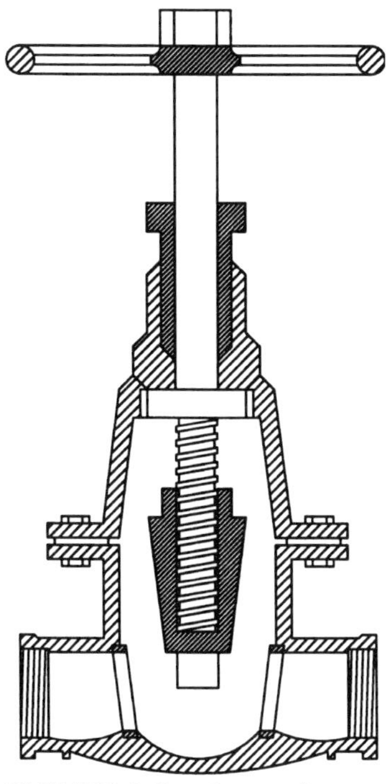

FIGURE 3.2 Gate valve

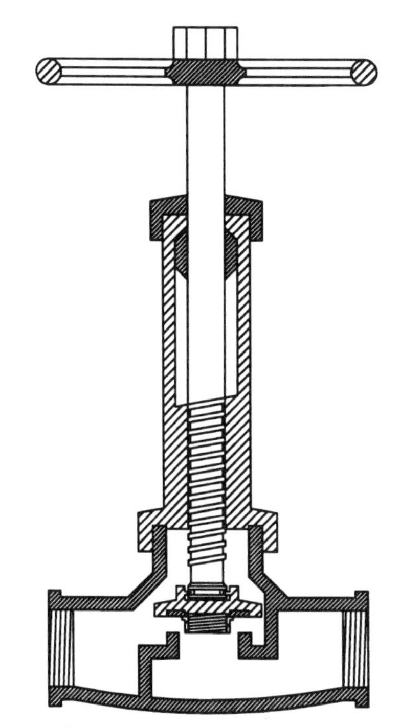

FIGURE 3.3 Globe valve

individual fixtures, branch supply lines that serve bathrooms and kitchens, and every riser (i.e., vertical supply line). A gate valve is the most commonly used device that can obstruct the flow of water by means of a gate-like wedge disk fitted within the valve body (Figure 3.2). It is used in locations where the valve is required to be left completely open most of the time. The gate valve mainly performs shut-off duty. It is not intended for flow regulations.

A globe valve is installed when it is necessary to regulate the flow of water. It is a compression type valve that controls the flow of water by means of a circular disk installed within the valve body (Figure 3.3). The globe valve usually has small ports, an "S" flow pattern, and relatively high pressure drop. Globe valves provide tight, dependable seals with minimum maintenance.

A check valve is used to prevent backflow in a supply system. It is used to direct the flow of water in only one direction. Any reversal of flow closes the valve (Figure 3.4).

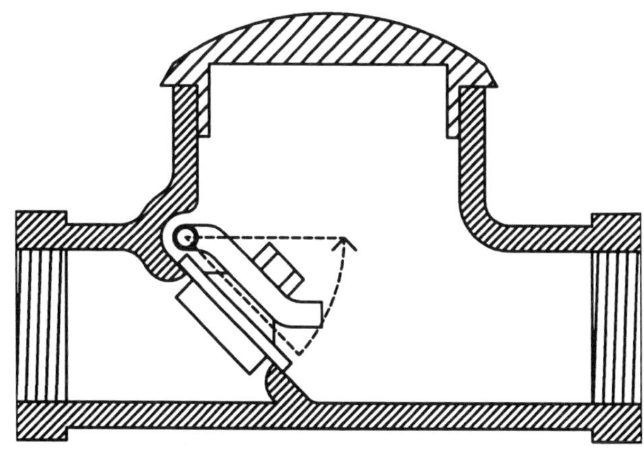

FIGURE 3.4 Check valve

Water Hammer Arrestor When a water supply valve or a fixture in a supply system is closed quickly, the force exerted by the fast-flowing water causes the pipe to shake and rattle. This is known as water hammer. It can be controlled by using a water hammer arrestor at the end of each fixture in the system.

Insulation of Pipes Pipes carrying cold water usually have a surface temperature of about 60°F. If the ambient air temperature rises to 85°F with a relative humidity of 40 percent, condensation will occur on the pipe surfaces. The higher the dry-bulb temperature and relative humidity, the more will be

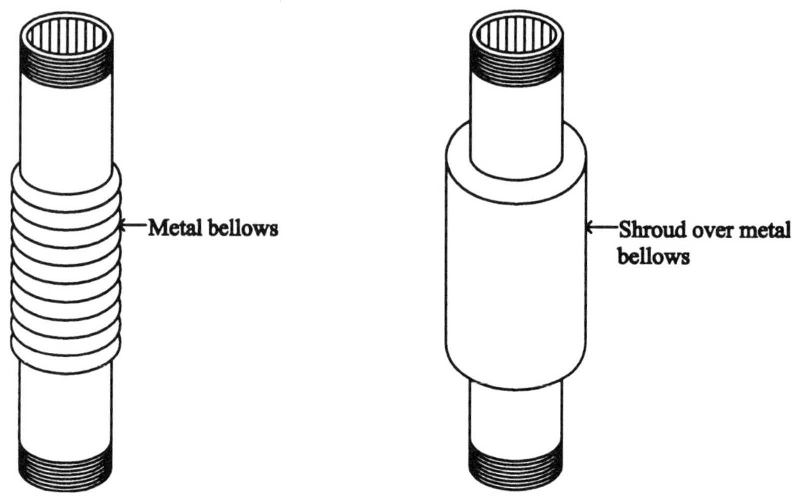

FIGURE 3.5 Bellows type expansion compensator

the condensation. Deterioration of finished surfaces will occur due to the dripping of the moisture from the supply pipes. All cold water pipes and fittings should be insulated to prevent condensation. Material commonly used for the purpose is fiberglass, 1/2 to 1 inch in thickness.

Hot water pipes should also be insulated using the same material to prevent heat loss to the surrounding air. Parallel hot and cold water pipes should be separated by a minimum of 6 inches to prevent heat interchange.

Pipe Expansion The temperature of water flowing through hot water supply pipes can range from 60°F to 180°F. It causes a significant temperature difference to exist between the indoor air and the hot water in supply pipes. Depending on the coefficient of expansion of the material, the piping system will experience some elongation. The actual expansion of pipe can be calculated using the following formula:

$$\text{Total expansion in feet} = L_0 \cdot \Delta t \cdot \alpha \qquad (4)$$

where L_0 = total pipe length in feet, Δt = temperature difference, and α = coefficient of expansion of the piping material.

Pipe expansion is negligible for small buildings, but may be quite significant for large and tall buildings. Expansion joints, as illustrated in Figure 3.5, are used to prevent pipes from buckling as a result of expansion.

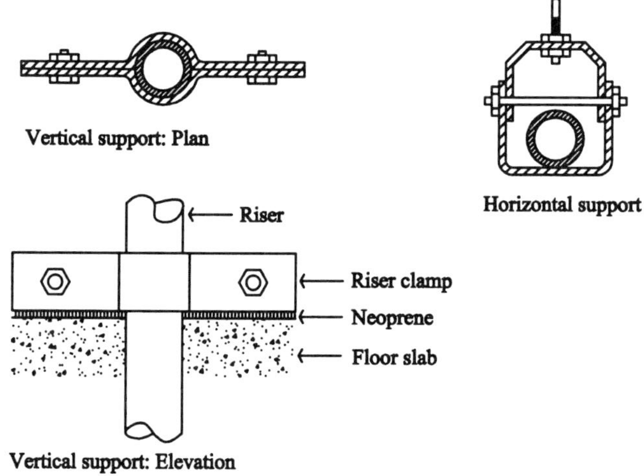

FIGURE 3.6 Pipe hangers

Pipe Support One cubic foot of water weighs 62.4 pounds. Water supply pipes are therefore quite heavy and require adequate support. Different types of anchors and hangers are used to provide support to the pipes (Figure 3.6). Horizontal pipes should be supported at 10-foot intervals. Closer spacing of 8 to 10 feet is preferred for metal pipes 1/2 inch and smaller in size. Plastic pipes 1 1/2 inch and smaller in size should be supported at 5-foot intervals. Vertical run of 1-inch pipes should be supported at every floor level; larger pipe sizes may be supported at every two floors.

PLUMBING FIXTURES

The plumbing fixtures are the visible part of a water supply system. They are the devices that receive water from a distribution system or that receive waterborne wastes and discharge them into the drainage system.

Minimum Requirements All nationally recognized plumbing codes normally include the minimum number of plumbing fixtures required to be installed in a building. The number is based on the use of the building and its population. Table 4.1 in Chapter 4 lists such requirements.

General Fixture Characteristics Fixtures are made of dense, impervious, and abrasion-resistant materials. Commonly used materials are enameled cast iron, stainless steel, fiberglass-reinforced plastic, and vitreous china. They are installed with fittings such as faucets or control valves, waste pipes, and other accessories as necessary.

Air Gap The connection between a water supply component and the relevant plumbing fixture must have an unobstructed vertical separation through the free atmosphere. This separation is known as an **air gap** (Figure 3.7). The purpose of maintaining this air gap is to prevent the possibility of contaminating supply water with wastewater. It is measured between the lowest opening from any pipe or faucet conveying potable water and the **flood-level rim** of the fixture. Minimum air gap is affected by the size of opening of a faucet and the side walls of a fixture. It ranges from 1 to 3 inches.

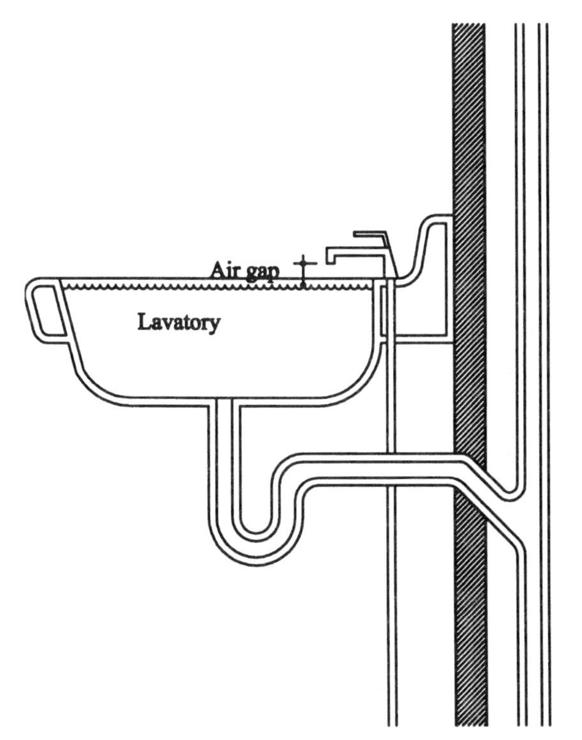

FIGURE 3.7 Air gap

Vacuum Breaker A vacuum breaker is installed at the branch connection to a plumbing fixture that lacks an air gap (e.g., dishwasher) in order to prevent backflow. If the water pressure in a supply system drops below the atmospheric pressure level, then it is possible for the foul water in the fixture to be siphoned into the supply system. Under such conditions, a vacuum beaker will close the pipe automatically in response to atmospheric pressure and thus prevent the backflow (Figure 3.8).

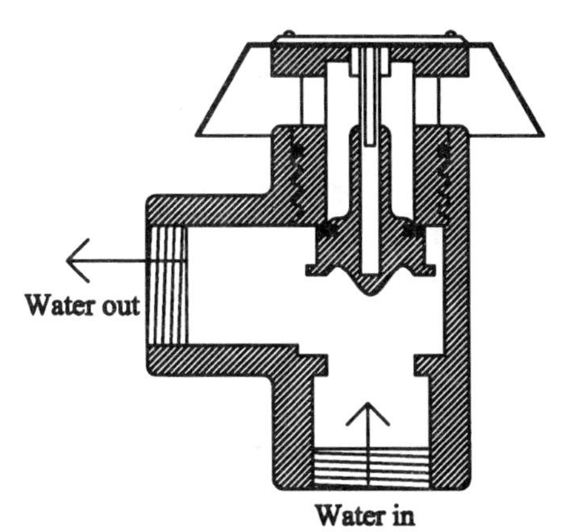

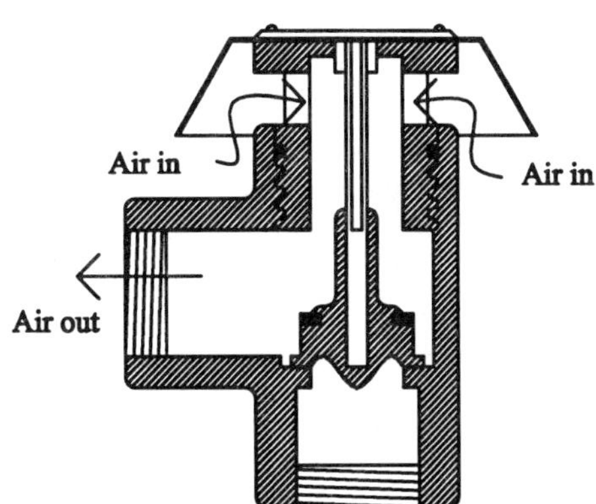

FIGURE 3.8 Vacuum breaker

Supply Fixture Units (SFU) Demand for water by a plumbing fixture varies according to its type and the occupancy category (i.e., either public or private) of the building in which it is installed. An index called supply fixture unit is used for measuring the demand load of a plumbing fixture. Table 4.2 in Chapter 4 provides the supply fixture ratings of standard plumbing fixtures. The table also lists the minimum water pressure required for proper operation of those fixtures. The supply fixture unit for a separate maximum demand for either cold or hot water supply by a fixture is 75 percent of the total supply fixture unit of that fixture.

SIZING OF SUPPLY PIPES

Total Water Demand In order to determine the size of water supply main to a building and the subsequent branch sizes, it is necessary to determine the maximum momentary load that the supply main should carry. This demand is calculated in terms of gallons of water per minute. Once the total supply fixture units for all the plumbing fixtures installed in a building have been calculated, the total water demand can be found out using one of the demand curves shown in Figure 4.4 of Chapter 4. These curves are based on a study by R. B. Hunter, who established a curvilinear correlation between the supply fixture units and the peak demand flow rate of the water supply distribution systems. These curves indicate that the demand rate does not increase proportionately to the increase in supply fixture units. This is because in most installations, the total number of plumbing fixtures is not expected to be used concurrently.

Curves labeled "valve" in Figure 4.4 (Chapter 4) are used for systems having predominantly flush-valve-operated water closets or urinals. Curves labeled "tank" are used for systems consisting of predominantly flush-tank-operated water closets or urinals. According to most building codes, all public buildings must use flush valves.

Water Pressure Sufficient pressure must be maintained in the building water supply system to ensure adequate flow of water through the pipes and the fixtures. The pressure components in a building supply system are the following:

- **Water pressure in municipal supply main (P_{SM}):** This is the available pressure at the point in the municipal supply main from which it is connected to the building supply main. This pressure may range from 50 to 70 psi. The local water supply authorities should be contacted to get the correct figure. The sum of the pressure losses in the distribution system described hereafter should be equal to the water pressure in the municipal supply main.
- **Pressure required for fixture (P_F):** This is the pressure needed to operate the most remote fixture on the highest floor of an upfeed distribution system. The figure can be obtained by consulting Table 4.3 in Chapter 4.
- **Pressure lost due to height (static pressure) (P_{HT}):** This is the pressure required to overcome the loss when water travels vertically from the level of the municipal supply main to the highest fixture in the building. It is a product of the total vertical distance traversed by water and unit static pressure.
- **Pressure loss due to flow of water through meter (P_M):** The water supply to a building is routed through a meter in order to measure the quantity of water consumed. The loss in flow through meter is a function of the demand rate and the size of the meter. This value is estimated, usually based on a meter size of 2 inches for residences and small offices. After the supply main size has been determined, the value must be rechecked and, if necessary, a recalculation should be made.
- **Pressure required for friction loss in piping (P_{FLH}):** Loss of pressure occurs because of the resistance offered to the flow of water by the pipe walls. It is also affected by the **total equivalent length** of the piping system. **Total equivalent length (TEL)** is the sum of **developed length (DL)** (i.e., the total linear distance of water travel from municipal main to the remotest and highest fixture) and **equivalent length (EL)** (i.e., length of the piping system equivalent to the fittings used for construction). The pipe length equivalent to the fittings is commonly estimated to be equal to 50 percent of the developed length. Therefore, the **TEL** of the supply piping of a building is equal to 1.5 times its **DL.** Total pressure loss in the distribution system is equal to the difference between

municipal supply main pressure and pressure losses generated by other pressure components in the system. Unit pressure loss due to friction, also known as friction loss in head, is calculated per 100 feet of total equivalent length. The following formula is used for the calculation:

- $P_{FLH} = [P_{SM} - (P_F + P_{HT} + P_M)] \cdot 100/TEL$ (5)

Sizing of Building Supply Main Once the demand rate in gallons per minute and friction loss in head per 100 feet of pipe length have been determined, the size of the building supply main can be obtained using the flow graphs shown in Figures 4.5 and 4.6 of Chapter 4. The horizontal lines in the chart represent demand rate, the vertical lines represent friction loss in head, the diagonal lines rising from left to right represent pipe sizes, and the diagonal lines rising from right to left represent the velocity of water flowing through the pipes. Examples of how to size supply pipes is given in Chapter 4.

Water Velocity Water flowing through supply pipes tends to produce noise due to friction. The higher the velocity, the greater the noise. Moving water can be heard within the pipes if the water velocity is higher than 10 feet per second. It may sometimes be necessary to install a higher pipe size in order to bring down the water velocity within this limit.

3.2 BUILDING DRAINAGE

BASIC ELEMENTS OF DRAINAGE SYSTEM

What comes in must go out; half of building plumbing is for getting rid of wastes. A drainage system consists of the least visible and least glamorous elements of a building's plumbing, but it is just as important as the supply system. It safely removes wastes for treatment and provides a critical barrier that keeps sewer gases, insects, and rodents from entering the building. Commonly known as DWV (drainage, waste, and venting), the system consists of drainage pipes, traps, and vents.

Drainage Pipes All plumbing fixtures receiving water from the supply system must discharge the used water to the drainage system through soil or waste pipes. **Soil pipes** convey wastes containing **fecal matter,** while **waste pipes** convey discharge free of any fecal matter. Vertical pipes are called either **soil** or **waste stacks.**

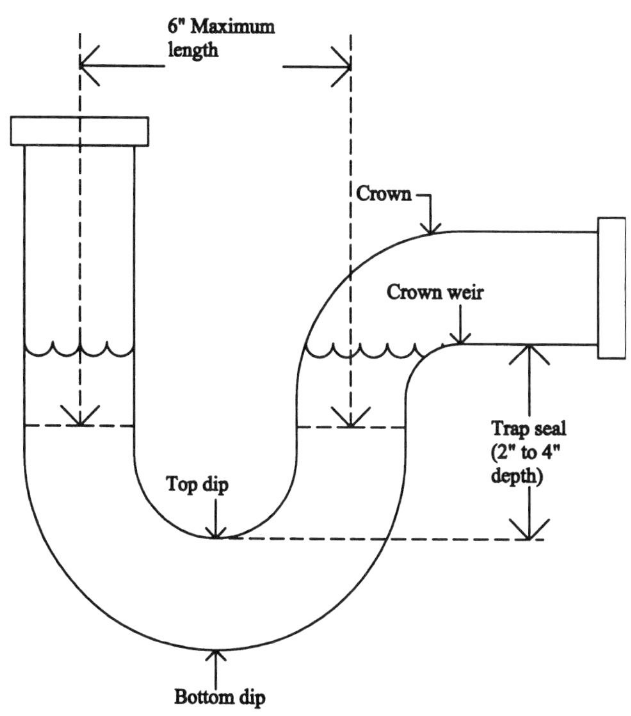

FIGURE 3.9 Fixture trap

Traps One of the basic principles of drainage is that every plumbing fixture must be installed with a trap. It provides a water barrier against the infiltration of sewer gases into the building without materially affecting the flow of sewage or wastewater through it (Figure 3.9). The water barrier in a trap is termed the trap seal, and may be defined as the column of water retained between the crown weir and the dip of the trap (Figure 3.9). The depth of a trap seal should not exceed a minimum of 2 inches and a maximum of 4 inches.

All plumbing fixtures must be equipped with a trap. They are usually placed within 2 feet of the fixture. The most common form of trap is constructed in the form of letter P; hence it is called a P-trap. All traps should be capable of being completely flushed each time a plumbing fixture is operated. Many are accessible for cleaning through a bottom opening closed by a plug.

Vents A drainage system will not function properly without the installation of vent pipes. Vents provide a free circulation of air within the drainage system.

They equalize pressure to aid drainage and allow sewer gases to escape to the atmosphere. Without venting, high pressure in the drains may force sewer gas out through traps and toilets. Also, low pressure in the drains may cause siphoning in the traps whenever the fixtures are drained. If the trap seals are broken, sewer gas will enter directly into the house (Figure 3.10).

A vertical pipe installed to provide circulation of air to and from the drainage system is called a **vent stack.** When a soil or waste stack is extended above the highest horizontal drain for a group of fixtures, it is called a **stack vent.**

In order to provide efficient ventilation to the drainage system, it is necessary to use individual or continuous vents. However, plumbing codes allow the use of circuit and loop vents, which do not provide an individual vent for every plumbing fixture.

A **circuit vent** serves more than two traps and extends from the front of the last fixture connection of a horizontal branch to the vent stack (Figure 3.11). This may be used on intermediate floors of multistoried building. A circuit vent that serves more than three water closets should have a relief vent in front of the first fixture that is connected to the circuit.

A **loop vent** is similar to a circuit vent except that it loops back and connects with a stack vent instead of a vent stack. Both these types of vents are dependent on the air space at the top of fixture branch pipes for delivery and removal of air from individual plumbing fixture connections. Therefore, the relevant branch drains should be conservatively sized.

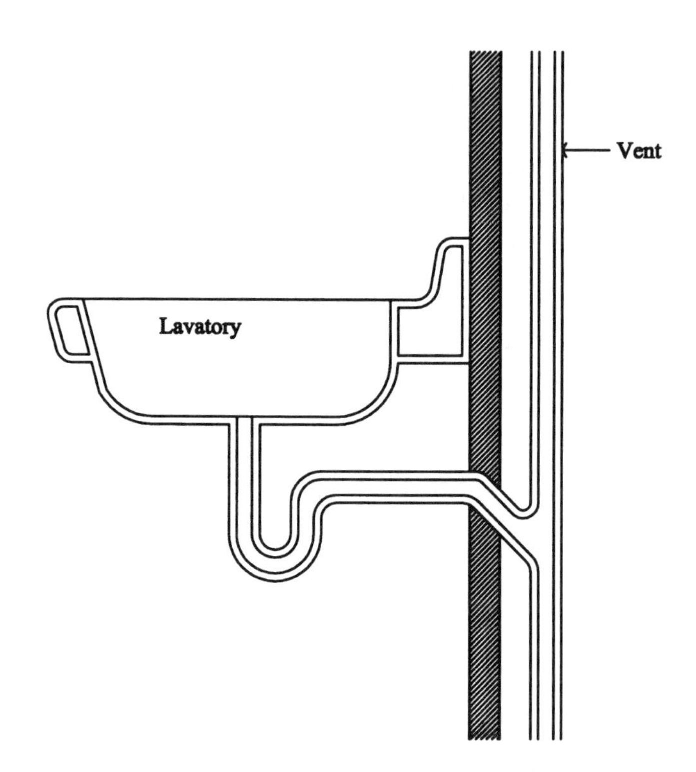

FIGURE 3.10 Fixture vent

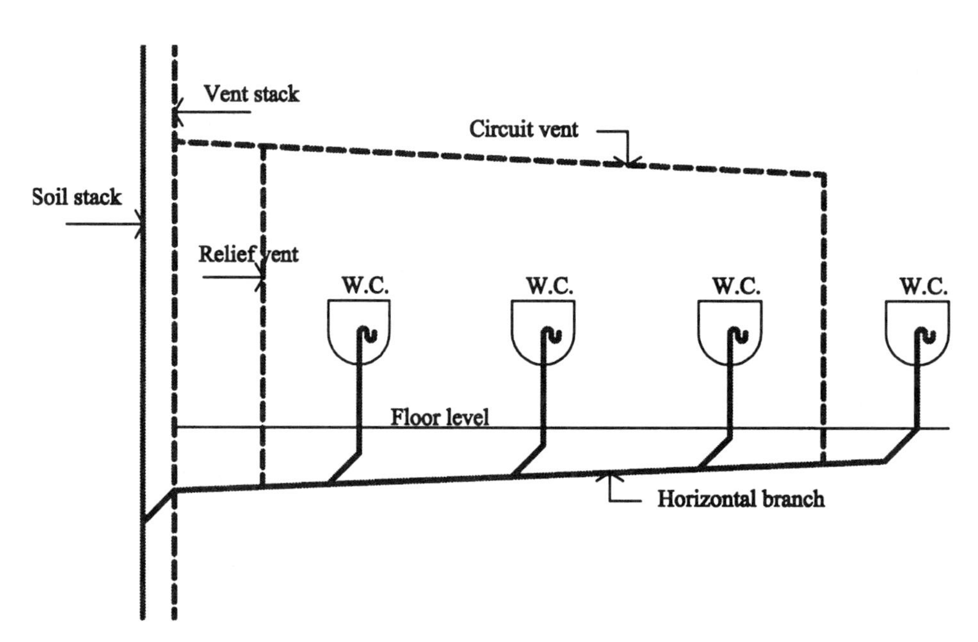

FIGURE 3.11 Circuit vent detail

Even though they are legal, the use of circuit and loop vents are not recommended in best practice.

PIPING AND FITTING MATERIALS

The principal materials used for piping different components of the drainage system are cast iron, copper, plastic, galvanized steel, and vitrified clay. Joining methods of pipes and fittings must be appropriate for the material used. Different types of joining methods are illustrated in Figure 3.12.

CHANGE OF DIRECTION

All direction changes in the flow line of a drainage system should be made with easy bends to avoid clogging of the pipes. Therefore, a branch drain should always be connected to a main drain at a 45° angle using a Y fitting or a combination of Y fitting and 1/8 bend (Figure 3.13). Right-angle connections are not used for the purpose.

CLEANOUTS

A **cleanout** is an accessible opening in a drainage system used for removal of obstructions. It is an essential part of the drainage system and should be provided in the following locations:

- at the outside wall of the building where the building drain connects to the house drain;
- at the base of all soil and waste stacks;
- at the upper terminal of all horizontal branch drains;
- every 50 feet on horizontal piping that is 4-inch size or smaller;
- every 100 feet on horizontal piping that is larger than 4 inches; and
- at all direction changes that are greater than 45°.

OTHER ACCESSORIES

There are many other accessories that are required for efficient operation and maintenance of a drainage system. Some of the common devices are described below.

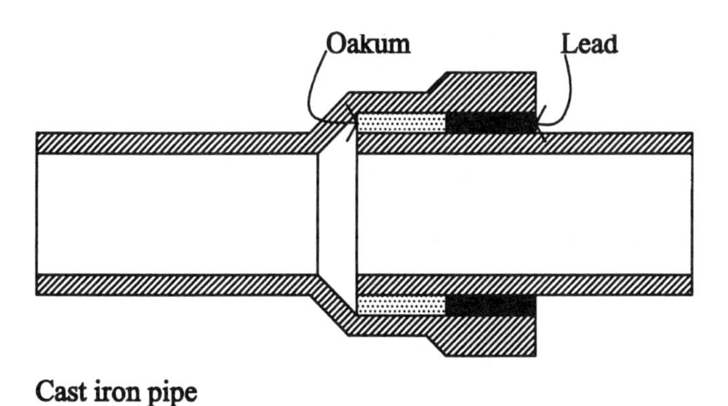

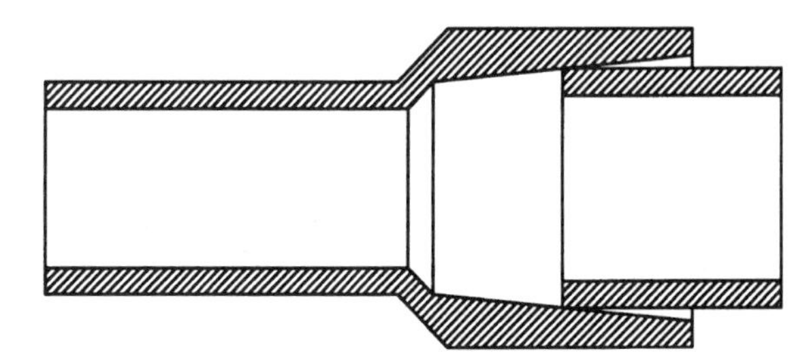

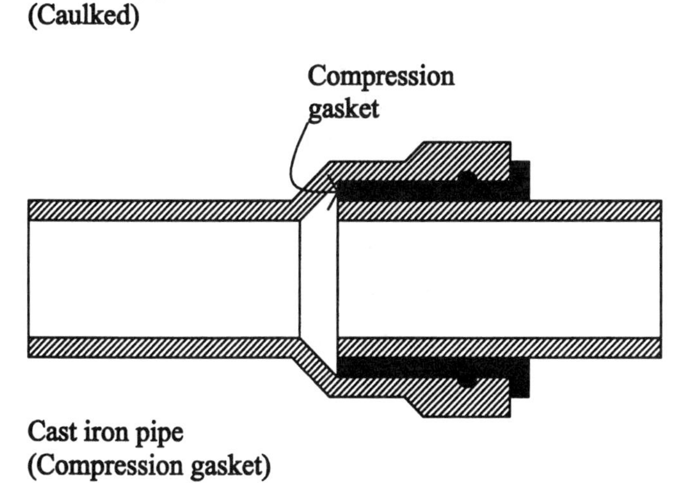

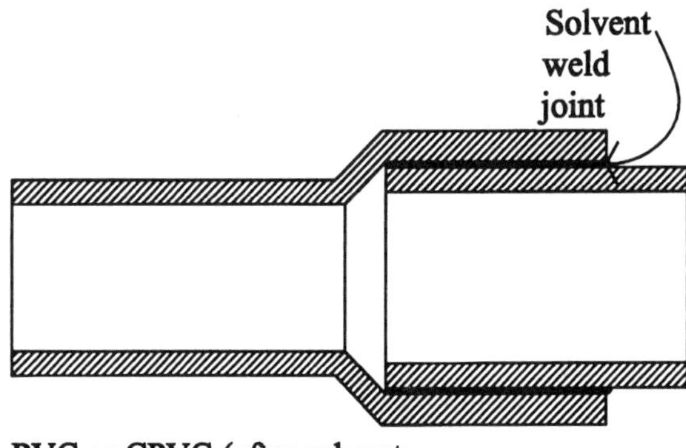

FIGURE 3.12 Pipe joints

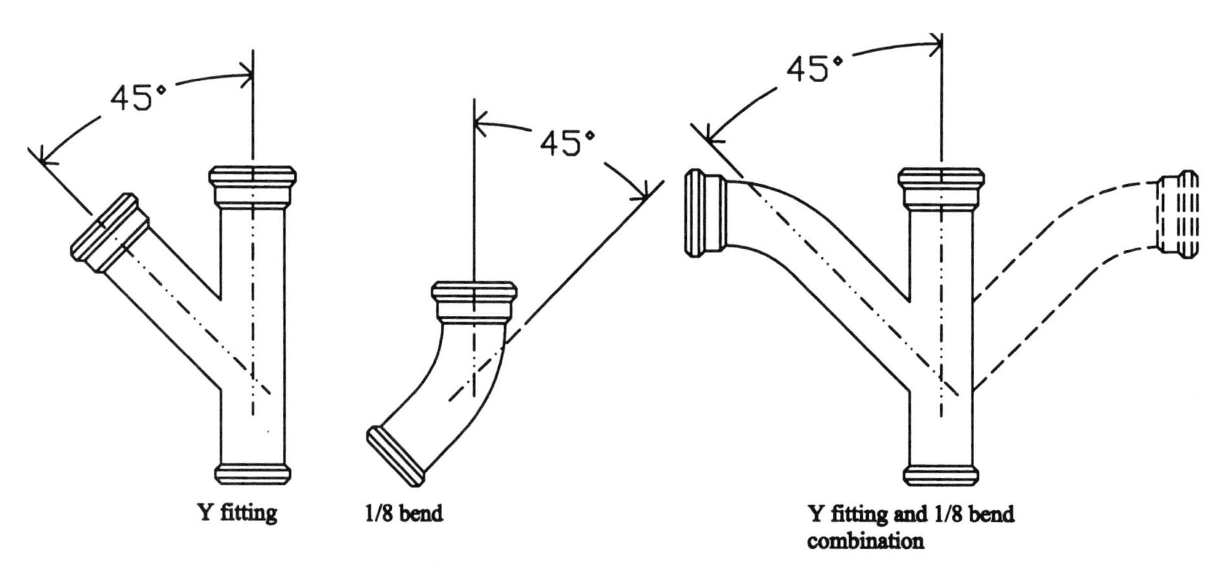

FIGURE 3.13 Drainage fittings

Floor Drain This is a receptacle installed on the floor to receive water that is to be drained from the floor into the drainage system. A floor drain is provided with a strainer and a trap, and is suitably flanged to provide a watertight joint in the floor. It is necessary in areas where floors are washed after food preparation and cooking, mechanical rooms, and rest rooms.

Backwater Valve This is a type of check valve installed to prevent the backflow of sewage from flooding the basement or lower levels of a building. It is not capable of protecting the entire drainage system, and should be used only when necessary.

Interceptor This is a device installed to separate and retain deleterious or hazardous matter from normal wastes, while allowing the normal wastes to discharge into the drainage system by gravity. It includes devices to intercept hair, grease, plaster, lubricating oil, glass grindings, or such other unwanted materials. A common interceptor installed in a residential drainage system is the grease trap.

SIZING OF DRAINAGE PIPES

Drainage pipes (mains and branches) are sized using an index called the **drainage fixture unit (DFU).** It is a measure of the probable discharge into the drainage system by various types of plumbing fixtures. The drainage fixture unit rating of a particular fixture depends on its rate of drainage discharge, on the time of duration of a single drainage operation, and on the average time between successive operations. One drainage fixture unit is approximately equal to 7 1/2 gallons of waste discharge per minute. DFU values of various plumbing fixtures are given in Table 4.3 in Chapter 4. Branch and main drainpipe sizes are selected directly from a table (see Tables 4.5, 4.6, and 4.7 in Chapter 4) in accordance with the load in DFU carried by the drain. Size of the pipes for individual fixtures are the same as their respective trap sizes. Following are some supplemental rules and recommendations for sizing of drains:

- No drain may be smaller than 1 1/4 inches.
- The maximum load on a 1 1/4-inch drain should not exceed one fixture unit.
- Codes allow a water closet to be connected to a 3-inch drain, but it is highly recommended to use a 4-inch minimum connection.
- A house drain may not be smaller than 4 inches.

An example of how to size drainage pipes is given in Chapter 4.

SLOPE OF HORIZONTAL DRAINS

Water and waste products discharged from plumbing fixtures into horizontal drainpipes flow under the pull of gravity. Therefore, these pipes have to be installed at a certain slope. It is expressed as the fall per foot of horizontal pipe length. Factors that affect slope or grade of drainpipe are pipe size and connected drainage fixture unit. Table 4.6 in Chapter 4 shows the relationship between slope and these factors.

SIZING OF VENT PIPES

Vent pipes are also sized using drainage fixture units. The other factors that affect the size of a vent pipe are the size of soil or waste stack that is required to be vented and the developed length of the vent. Table 4.5 in Chapter 4 may be used for determining vent sizes for a drainage system. Following are some supplemental rules and recommendations for sizing of vents:

- No vent may be smaller than 1 1/4 inches.
- An individual fixture vent should not be smaller than half the fixture drain size.
- An individual fixture vent should not be larger than the fixture drain.
- Loop and circuit vents are usually allowed by code, but not recommended in best practice.
- A water closet must have a 2-inch vent.
- A vent stack or stack vent that penetrates a roof must be of a size not less than 4 inches and must rise 12 inches above the roof.

An example of how to size vents is given in Chapter 4.

SINGLE-PIPE DRAINAGE SYSTEM

A single-pipe drainage system consists of only drainpipes. It is a specially engineered method, known as the **sovent system,** suitable for high-rise buildings. It is an all copper or plastic system consisting of an aerator fitting at each floor and a de-aerator fitting at the bottom of each stack. The effluent from the plumbing fixtures is mixed with air in the aerators to produce a foamy substance that lacks the stack-filling tendency of the liquid effluent. It does not produce pressures of more than 1 inch water gauge. Therefore, a normal trap seal is safe against being siphoned out or penetrated. It eliminates the need for vent pipes.

3.3 WATER CONSERVATION

The United States, in general, has abundant water resources. But some parts of the country, particularly the western region, have an emerging problem with water supplies adequate to meet the current consumption patterns. The rate of consumption of groundwater exceeds the rate at which it is replenished by recharge. Recent drought in some areas has accentuated the need to balance water demand with available supply. The future health and economic welfare of the nation's population are dependent upon a continuing supply of fresh, uncontaminated water.

One of the ways to maintain this balance is through water conservation. Simply stated, water conservation means doing the same with less, by using water more efficiently or reducing where appropriate, in order to protect the resource now, and for the future.

Using water wisely will reduce pollution and health risks, lower water costs, and extend the useful life of existing supply and waste treatment facilities. In order to promote the conservation and efficient use of water, the Energy Policy and Conservation Act was enacted in the United States in 1992. The statute establishes national water conservation standards for:

- Showerheads—2.5 gallons per minute
- Toilets—1.6 gallons per flush
- Faucets—2.5 gallons per minute
- Urinals—1.0 gallon per flush

Over 42 percent of our indoor water use is for flushing water closets. The conventional toilet uses at least 3.5 gallons per flush. If low-flush toilets using 1.6 gallons per flush replace them, the overall indoor use is reduced by over 20 percent. The low-flush toilets are proven technology. They work efficiently and without problem.

The shower accounts for approximately 20 percent of indoor water use. The typical showerhead allows a water flow of 4 to 8 gallons per minute. Installing a low-flow showerhead with a flow rate of 2.5 gallons per minute will reduce consumption by one-half, yet most people will not notice the effects of the reduction.

Water-conserving faucet aerators are available in sizes ranging from approximately 0.5 to 2.5 gallons per minute. Low-flow aerators mix air with the water to make an effective spray pattern. Therefore, by installing a low-flow aerator, one can save a lot of water.

Washing clothes accounts for approximately 25 percent of residential water use. Front-loading horizontal-axis machines use one-third less water than top-loading vertical-access machines. The standard top-loader uses 35–55 gallons per load, whereas a front loader will use 25–30 gallons per load. As well as saving water, the front-loading machines also save energy. Front-loading machines, however, are still more expensive than the standard top-loading models in the United States.

REVIEW QUESTIONS

1. The recovery rate of a water heater is indicated to be 5 gph. What does it mean?
2. Water pressure in a municipal street main is 60 psi. How many stories (where floor-to-floor height is 10 feet) can a building have in order to receive water supply on all the floors? Assume that a minimum pressure of 25 psi is required to operate all the plumbing fixtures on the top floor.
3. What type of plastic piping would you recommend for hot water supply?
4. What type of copper tube has the heaviest wall?
5. What does a check valve do?
6. It is recommended that the supply pipes should be insulated. Why?
7. What is a trap?
8. What is the minimum recommended depth of a trap seal?
9. What is a vent stack?
10. What is a drainage fixture unit?

ANSWERS

1. It indicates that the water heater is capable of producing hot water, at a specified temperature, at the rate of 5 gallons per hour.
2. Pressure available to traverse vertical distance = 60 psi − 25 psi = 35 psi
   Vertical distance traversed by water having a pressure of 35 psi = 35/0.433 = 80 feet
   Number of floors of the building = 80/10 = 8
3. Polyvinyl dichloride
4. K-type
5. It prevents backflow in a water supply system.
6. To prevent condensation on pipe surfaces; to prevent heat exchange between supply water and the surrounding air.
7. It is a fitting, installed in a drainage system, that provides a water barrier against the infiltration of sewer gases into a building.
8. 2 inches
9. It is a vertical pipe that provides circulation of air to and from a drainage system.
10. It is an indication of the probable discharge of waste by a plumbing fixture into the drainage system.

Plumbing Examples

hot water left~cold water to the right, shake hands judiciously

This chapter begins with a discussion of the sequential plumbing activities that meet a construction schedule. Architectural drawings locate plumbing fixtures with appropriate clearances for accessibility, and plumbing drawings detail the piping and connections needed to serve each fixture.

Next a series of tables and graphs explains the estimates and calculations involved in sizing a building's plumbing system. Complete the examples that follow each table or graph to develop your ability to number and size plumbing components.

Two example buildings, a 1,600-sq. ft. home and a 2 story 22,000-sq. ft. office building are used to illustrate supply and DWV piping layouts. Plans and drawings define the components of plumbing installations for both examples, and high-rise plumbing system components are also discussed. Review these examples in detail, and then prepare plumbing drawings for a building you have designed. Be sure to verify fixture requirements and plumbing tables in the applicable *local* building and plumbing codes. Tables here have been developed from various codes and will NOT apply in all localities.

4.0 PLUMBING SEQUENCE

Three sequential plumbing activities must fit the construction schedule. Underground work begins after the building corners are set, and must be completed before work on the floor slab or floor framing begins. "Rough in" work begins after the wall framing and roof deck are in place, and must be completed before insulation and sheetrock work can start. "Set and finish" work can begin when interior sheetrock, tile, and base cabinets are complete.

UNDERGROUND

Piping below the ground floor (Figure 4.1) is installed and capped before placing concrete or setting floor joists. Drain lines are tested by filling the pipe with water and looking for leaks before backfilling trenches. Water supply piping with underground joints and fittings is pressure tested with air or water to reveal possible leaks. Tests are not usually required for loop copper water supplies because all joints will be made above the floor.

Drain piping below the slab falls 1/8″ or 1/4″ per foot, so long runs to the city sewer can require deep trenches. All site piping is kept at least 18″ below grade, and water supply lines must be set below the frost line. Site trenches should be filled with sand prior to backfill to protect pipes and warn future excavators.

Natural gas piping is NOT run below the ground floor, for leaks could follow the pipe into the building.

ROUGH IN

When the roof deck and wall sheathing are complete, the plumbing crew "roughs" for fixtures and hose bibs. Water supply pipes and drainpipes are stubbed at the proper height for individual fixtures, and DWV trees are completed through the roof (Figure 4.2).

In high-rise buildings, DWV trees for each floor can be shop fabricated to minimize site labor. Horizontal vent branches are sloped to drain, and the plumbing contractor provides flashings for vents that penetrate the roof. Codes require at least one large VTR (vent through roof) and thicker walls are used behind water closets to enclose large vent and soil stacks. Wherever possible, small individual vents are

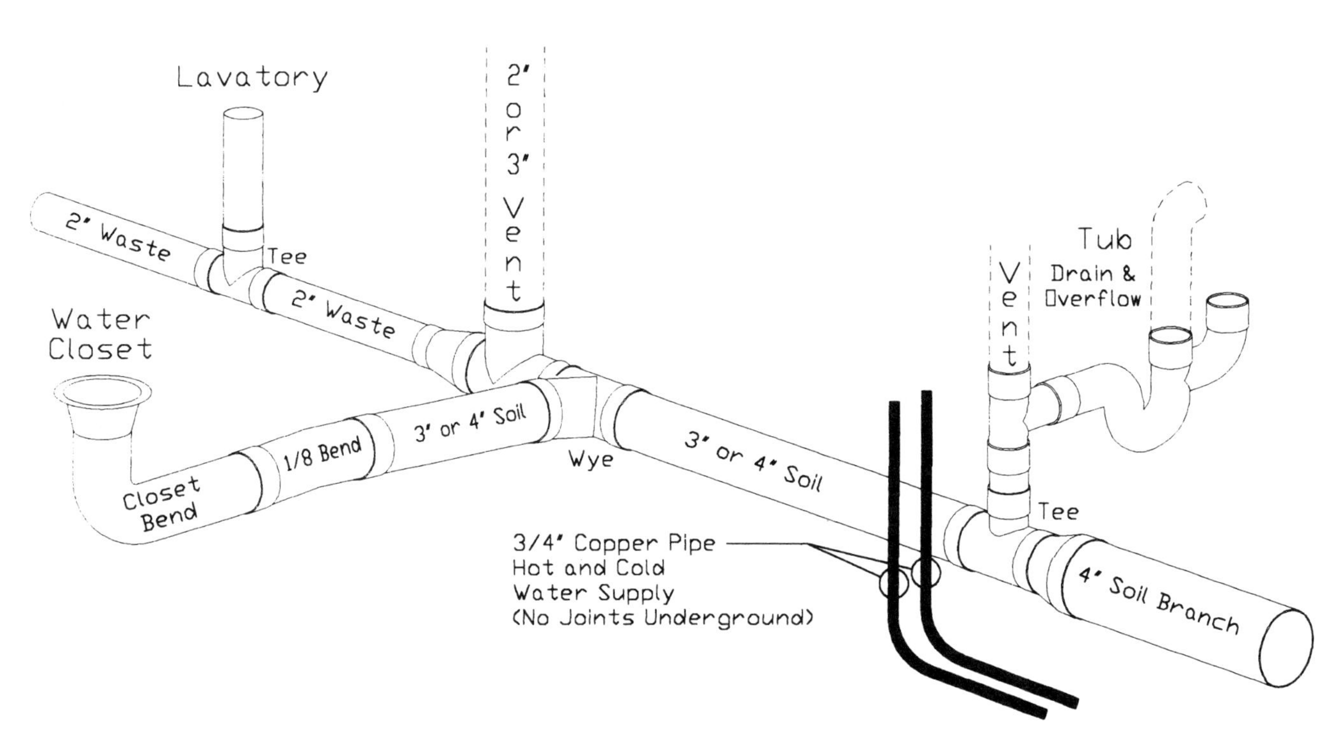

FIGURE 4.1 Underground piping schematic

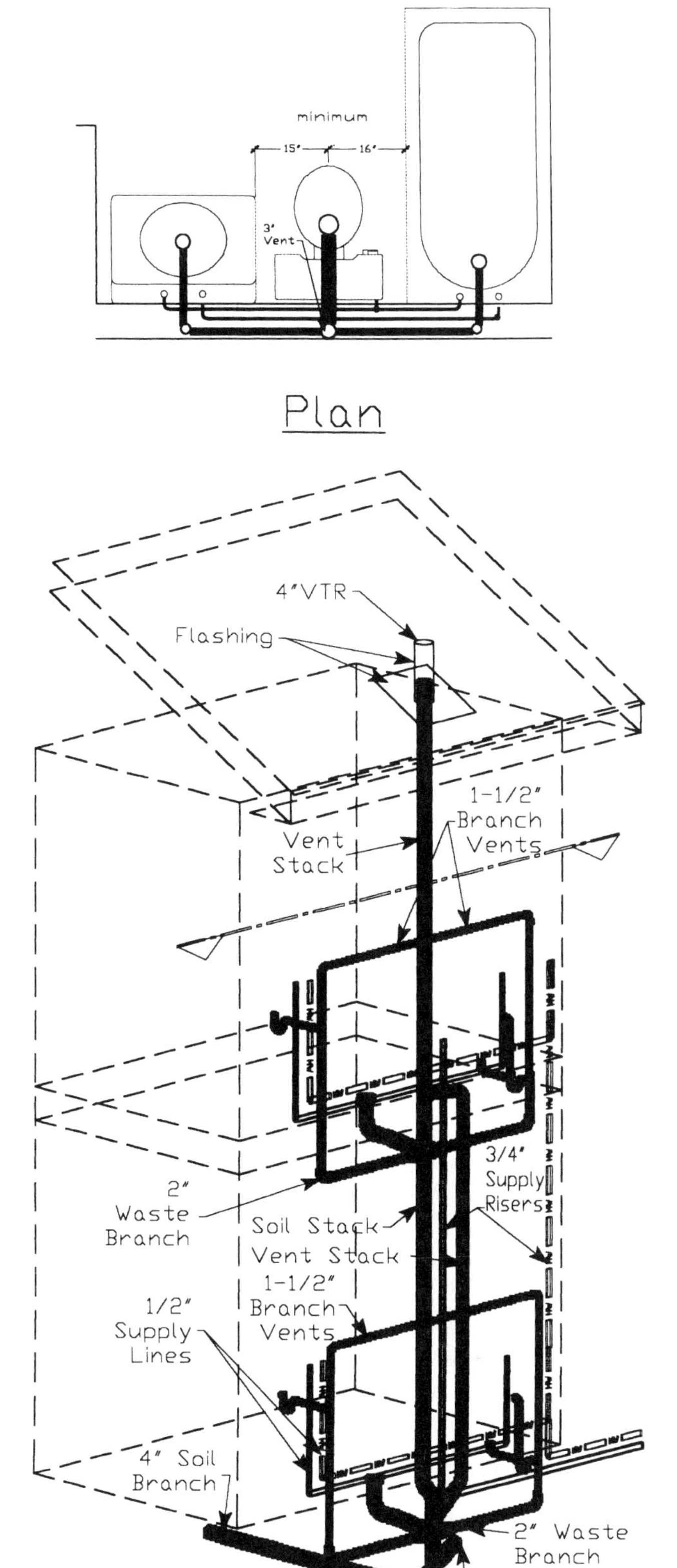

FIGURE 4.2 Plumbing Tree

connected to large vents in the wall or above the ceiling to minimize floor and roof penetrations.

Built-in fixtures like one-piece tub-showers are set now, but sinks and toilets are not installed until tile work, sheetrock, and counters are complete.

Water piping in vented roofs or attics is insulated to prevent freezing, and in very cold climates drain valves or freeze-proof hose bibs are specified. Black steel pipe is used for interior gas piping because natural gas reacts with galvanized steel.

When "rough in" is complete, piping is leak tested again, before building insulation and sheetrock work begins. DWV trees are capped and filled with water, hot and cold water supply lines are checked for leaks under pressure, and natural gas piping must maintain air pressure for 24 hours.

After "rough in," the building insulation contractor seals plumbing and electrical penetrations with expansive foam to minimize air infiltration.

SET AND FINISH

When interior walls, ceilings, and cabinets are complete, the plumbing crew returns to set toilets, sinks, water heaters, and other fixtures, and to install the valves and trim (Figure 4.3).

After final inspection, connections are completed to the city water and sewer mains, supply piping is flushed with chlorinated water, and all fixtures are checked for proper operation.

4.1 PLUMBING REQUIREMENTS

Local codes set minimum requirements for building plumbing systems. Most cities and towns adopt one of the National Codes, and then modify it to suit particular local requirements. A single National consensus code is planned for the year 2000.

Work the examples on each of the following pages to develop your understanding of plumbing system requirements.

Use Table 4.1 to work the following examples before checking your answers.

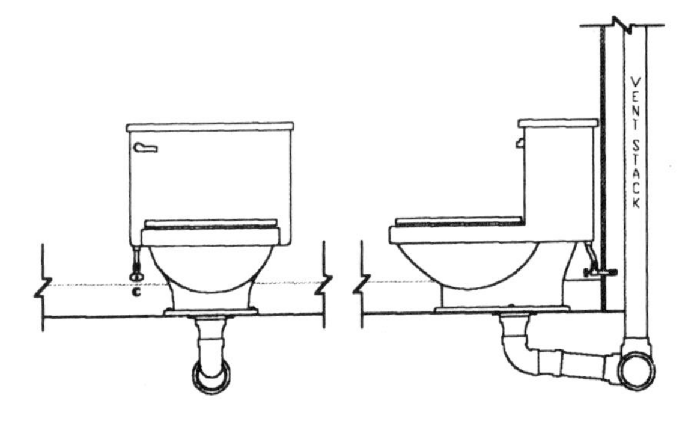

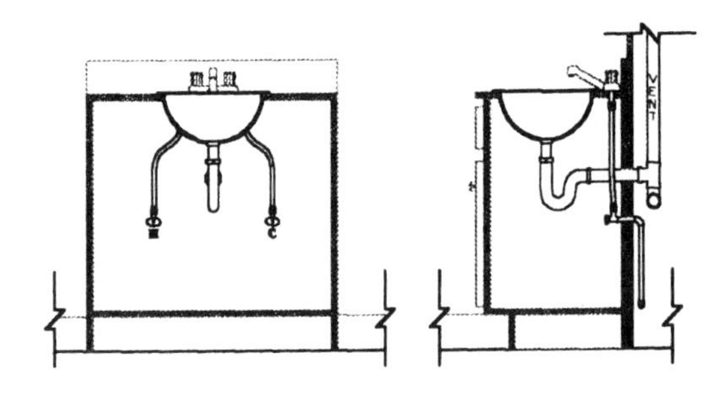

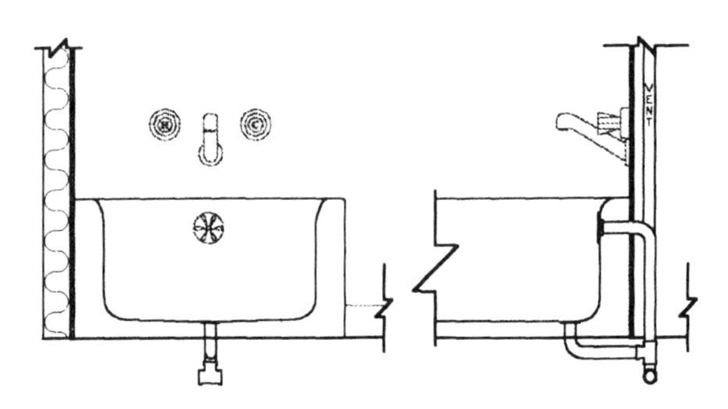

FIGURE 4.3 Setting fixtures, valves, and trim

TABLE 4.1

Occupancy Estimates

WARNING! Use the following APPROXIMATE values ONLY for first estimates. Codes vary and many exceptions apply. *Local codes govern.*

Floor Area In Square Feet Per Occupant

Occupancy	Floor Area
Church - Theater - Restaurant - Stadium - Arena etc.	
-fixed chairs—allow 1 occupant per seat	
-fixed pews or benches—(allow 18″ per seat)	
-moveable chairs	7 sq. ft.
-moveable chairs and tables	15 sq. ft.
-standing room only	3 sq. ft.
Office	*150 sq. ft.
Retail ground floor and basement	30 sq. ft.
-other floors	60 sq. ft.
School - classroom	20 sq. ft.
Library - reading areas	*50 sq. ft.
-stack areas	100 sq. ft.
Hospital	*240 sq. ft.
Residential	*200 sq. ft.

**Gross* building floor area enclosed by exterior walls. Values without asterisk are *Net* occupied room areas.

EXAMPLES

1. An office building has a gross floor area of 44,000 sq. ft. Estimate occupancy.
2. A church sanctuary has 200 pews, each 12′ long. Estimate occupancy.
3. A 4,400 sq. ft. night club includes 4,000 sq. ft. with moveable chairs and tables, plus 400 sq. ft. of standing room. Estimate the number of occupants.
4. A school includes ten classrooms; each is 600 sq. ft. Estimate occupancy.
5. A retail store has a total floor area of 15,000 sq. ft. Estimate occupants.
6. A 10,000 sq. ft. library includes 4,000 sq. ft. of reading area and 6,000 sq. ft. of stacks. Estimate occupants.

ANSWERS

1. Office = *293* — 44,000 ÷150
2. Church = *1,600* — (200)(12) ÷ 1.5
3. Night club = *400* — (4,000 ÷ 15) + (400 ÷ 3)
4. School = *300* — (600)(10) ÷ 20
5. Grocery store = *500* — 15,000 ÷ 30
6. Library = *140* — (4,000 ÷ 50) + (6,000 ÷ 100)

HOW MANY FIXTURES?

Building codes use occupancy estimates to set the required minimum number of WC's (water closets), Lav's (lavatories), and DF's (drinking fountains). The following brief tables are satisfactory for preliminary estimates, but verify fixture requirements with the applicable *local* code authority before preparing contract drawings.

When the building occupants include equal numbers of males and females, codes allow for 60 percent of total occupancy by each gender.

TABLE 4.2

Fixtures Required (minimum)

Private	*WC*	*Lav*	*Bath*
Hotel/Motel - room	1	1	1
Hospital - private room	1	1	1
Residence	1	1	1

WARNING! Use the following APPROXIMATE values ONLY for first estimates. Codes vary and many exceptions apply. *Local codes govern.*

Public	*WC**	*Lav*	*DF*
maximum number of occupants per fixture			
Church - theater			
-first 50	50	1/2 WC	100
-over 50	150	1/2 WC	100
Dormitory			
8 dorm occupants per shower			
Office & Retail			
-first 15	15	1/2 WC	100
-over 15	25	1/2 WC	100
Educational - nursery	15	1/2 WC	30
-elementary school	25	1/2 WC	40
-secondary school	30	1/2 WC	50
-college*	40/30*	1/2 WC	100
Hospital Ward	8	1/2 WC	100
20 patients/bath			
occupants per fixture: male—female			
College, male/female	40/30	1/2 WC	100
Restaurant			
-first 150, male or female	50	1/2 WC	200
-over 150, male/female	200/100	1/2 WC	200
Night club			
-first 40, male or female	40	1/2 WC	75
-over 40, male/female	40/20	1/2 WC	75
Stadiums & arenas			
-first 150, male or female	50	1/2 WC	200
-over 150, male/female	300/150	1/2 WC	200

*Up to 50% of required male WC's may be replaced one-for-one with urinals.

EXAMPLE ESTIMATES

Estimate the number of water closets, lavatories, urinals, and drinking fountains required in a one-story 44,000- sq. ft. office building.

Solution Estimate 293 building occupants (refer to preceding example #1). Estimate 60% men and 60% women, say 176 of each (if one gender dominates, adjust the percentages of males and females as appropriate).

- Eight WC's are required to serve 176 women (1 for the first 15, plus 6.44 for the next 161 @ 1 for 25 5 7.44; say 8).
- Eight more WC's are required to serve 176 men. Conserve space by replacing half these WC's with urinals.
- Eight lavatories are required in the building—half of the total WC requirement of 16.
- Three drinking fountains are required to serve 293 occupants at 1 DF per 100. In multistory buildings, codes also require at least one DF on *each floor*.

ANSWER

Minimum fixture requirements include:
12 WC's, 4 urinals, 8 lavatories, and 3 DF's.

MORE EXAMPLES

Use Table 4.2 to estimate the minimum number of WC's, urinals, lavatories, and DF's required to serve the following occupancies. Make your estimates before checking the solutions in the column at right.

1. A Theater* seating 1,600
2. A 4,400 sq. ft. night club for 400
3. An elementary school classroom wing for 600 students
4. A store with a total of 500 customers and employees
5. A library for 140

ANSWERS

1. A theater* seating 1,600 requires:
   estimate 960 of each sex (60%)
   WC — 1 for first 50 plus 6 for the remaining 810, or *7* for each sex = 14.
   UR — replace 50% = 3 men's WC.
   LAV — half of required WC = 7.
   DF — 1 per 100 = 16.
   WC = 11, UR = 3, LAV = 7, DF = 16
2. A night club for 400 occupants requires:
   estimate 240 of each sex (60%)
   WC — 1 for first 40 men and women, plus 5 for the remaining 200 men, plus 10 for the remaining 200 women = 17.
   UR — replace up to 50% = 3 men's WC.
   LAV half of required WC = 9.
   DF — 1 per 75 = 5.33, say 6.
   WC = 14, UR = 3, LAV = 9, DF = 6
3. An elementary school classroom wing for 600 students requires:
   estimate 360 boys and 360 girls
   WC — 1 for 25 = 30, half for boys and half for girls.
   UR — replace 50% = 7 boys' WC.
   LAV — half of required WC = 15.
   DF — 1 per 40 (600÷40) = 15
   WC = 23, UR = 6, LAV = 15, DF = 15
4. A store serving 500 requires:
   allow 300 men and 300 women (60%)
   WC — 1 for first 15, plus 12 for the remaining 285 of each sex = 26.
   UR — replace up to 50% = 6 men's WC.
   LAV — half of required WC = 13.
   DF — 1 per 100 = 5.
   WC = 20, UR = 5, LAV = 13, DF = 5
5. A library for 140 requires:
   estimate 84 men and 84 women (60%)
   use the table values for "college"
   WC — 1 for each 40 men = 3, plus 1 for each 30 women = 3, total = 6.
   UR — replace 50% = 1 men's WC.
   LAV — half of required WC = 3.
   DF — 1 per 100 = 1.4 say 2
   WC = 5, UR = 1, LAV = 3, DF = 2

*In some commercial and institutional occupancies like theaters and prisons, facilities for employees (actors or guards) must be calculated separately.

4.2 FIXTURE UNITS

Fixture units are numbers that indicate water quantity. A *SFU* (supply fixture unit) is a flow of about 1 GPM, and a *DFU* (drainage fixture unit) is a flow of about 0.5 gpm. Fixture units are used to determine pipe sizes.

TABLE 4.3

Fixture Units

Private	*SFU*	*DFU*	*psi**
Bathroom group (gravity tank)	6	6	10
Bathroom group (pressure tank)	5	5	25
Bathroom group (flush valve)	8	8	25
Lavatory	1	1	10
Tub or shower	2	2	10
Water closet (gravity tank)	3	4	10
Water closet (pressure tank)	2	2	25
Water closet (flush valve)	6	6	25
Kitchen sink	2	2	10
Washer (clothes—8 lb.)	2	3	10
Dishwasher	1	2	10
Hose bib	4	—	10+
Public	***SFU***	***DFU***	***psi***
Lavatory	2	1	10
Tub or shower	4	2	10
Urinal (gravity tank)	3	2	10
Urinal (flush valve)	5	4	15
Water closet (gravity tank)	5	4	10
Water closet (pressure tank)	2	2	25
Water closet (flush valve)	10	6	25
Kitchen sink	4	3	10
Service sink	3	3	10
Drinking fountain	1/4	1/2	10
Hose bib	4	—	10+

psi = minimum fixture supply pressure

EXAMPLE

Use Table 4.3 to calculate SFU and DFU for a public building with 12 water closets, 4 urinals, 8 lavatories, and 3 drinking fountains.

ANSWER

Supply		SFU
Public, flush valve WC's — 12@ 10	=	120
Public, flush valve urinals — 4@ 5	=	20
Public lavatories 8@ 2	=	16
DF's 3@1/4 say	=	1
Hose bibs, allow for 6@ 4	=	24
Total SFU	=	181

Drainage		DFU
Public, flush valve WC's — 12@ 6	=	72
Public, flush valve urinals — 4@ 4	=	16
Public lavatories 8@ 1	=	8
DF's 3@1/2, say		2
Total DFU	=	98

Answer: 181 SFU and 98 DFU

EXAMPLES

Calculate total SFU and DFU for the following:

1. A *hotel* wing with 400 private bathrooms using pressure tank water closets.
2. A 100 unit *apartment* with 25% 1 BR, 50% 2 BR, and 25% 3 BR. One-bedroom units have 1 bathroom, 2- and 3-bedroom units have 2 bathrooms, and all water closets have gravity tanks.
3. A *school* with 23 flush valve water closets, 6 urinals, 15 lavatories, 2 service sinks, 15 drinking fountains, and 6 hose bibs.

ANSWERS

1. The hotel has 2,000 SFU and 2,000 DFU.

Supply	SFU
Private bathroom group 400 @ 5	=2,000

Drainage	DFU
Private bathroom group 400 @ 5	=2,000

SFU and DFU for drinking fountains, hose bibs, service sinks, etc., would be added for a typical hotel wing.

2. The apartment has 1,666 SFU and 1,250 DFU.

Supply		SFU
175 bathroom groups @ 6	=	1,050
100 kitchen sinks @ 2	=	200
Washroom, 8 clothes washers @ 2	=	16
allow 100 hose bibs @ 4	=	400*
Total SFU	=	1,666

Drainage		DFU
200 bathroom groups @ 6	=	1,050
100 kitchen sinks @ 2	=	200
Total DFU	=	1,250

3. The school has 324 SFU and 191 DFU.

Supply		SFU
Public, flush valve WC's — 23@ 10	=	230
Public, flush valve Urinals — 6@ 5	=	30
Public Lavatories — 15 @ 2	=	30
service sinks — 2 @ 3	=	6
DF's 15 @ 1/4 say	=	4
hose bibs — 6 @ 4	=	24
Total SFU	=	324

Drainage		SFU
Public, flush valve WC's — 23@ 6	=	138
Public, flush valve Urinals — 6@ 4	=	24
Public Lavatories — 15 @ 1	=	15
Service sinks — 2 @ 3	=	6
DF's 15 @ 1/2 say	=	8
Total SFU	=	191

*When a landscape sprinkler system is used, include its sfu and reduce the hose bib allowance (diversity).

SUPPLY GPM

The graphs in Figure 4.4, p. 58 link total SFU to water supply demand in gallons per minute (supply GPM). Read solid-line curves for residential and commercial occupancies; use the dashed curves for large-assembly occupancies and supply risers in multiriser occupancies.

EXAMPLES

1. A *hotel* wing uses pressure tank water closets in each room. Find GPM demand for a supply riser that serves 2,000 SFU.
2. A 100-unit *apartment* with gravity tank water closets has a total of 1,666 SFU. Find GPM demand.
3. A school with flush valve water closets has 324 SFU. Find GPM demand.

*Large-assembly examples: stadium, theater, arena.

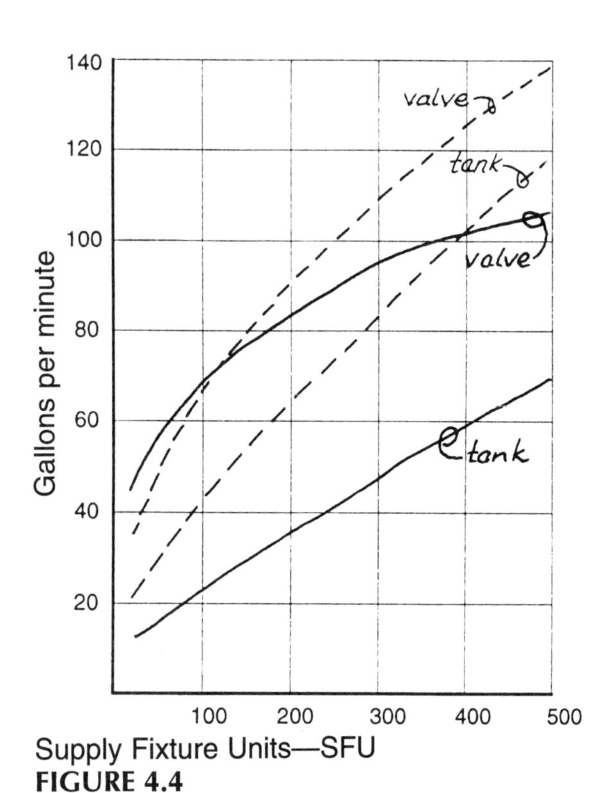

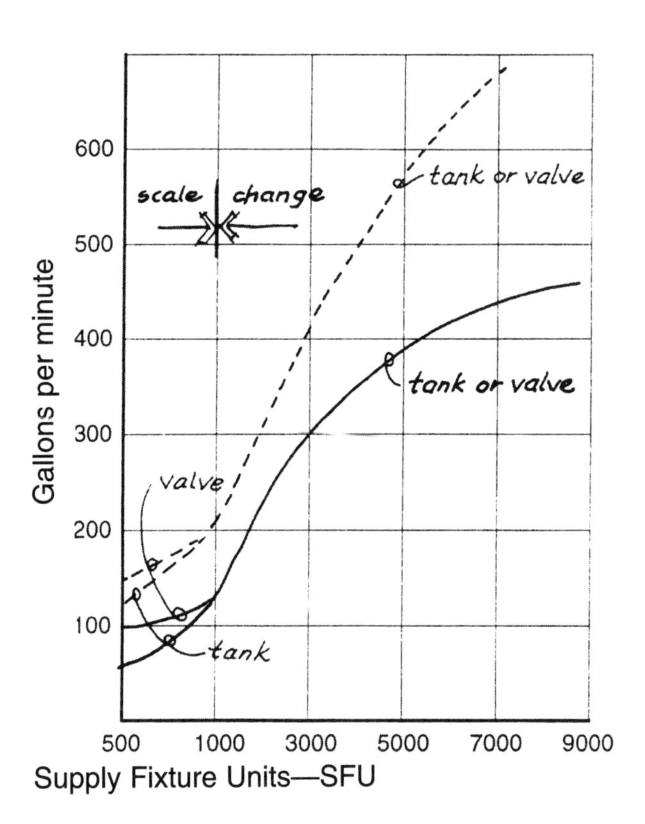

FIGURE 4.4

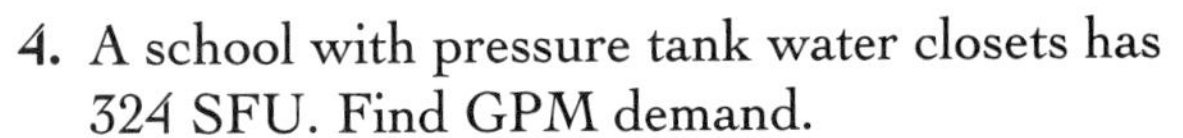

4. A school with pressure tank water closets has 324 SFU. Find GPM demand.
5. A theater has 400 SFU. Estimate demand GPM with flush valve water closets.
6. A stadium has 3,000 SFU. Estimate demand GPM with flush valve water closets.

ANSWERS

1. The hotel is a multiriser occupancy, so read the dashed curve. With 2,000 SFU, the demand is about 310 GPM.
2. Read the solid-line curve for this residential occupancy. With 1,666 SFU, the demand is about 200 GPM.
3. Read the solid-line curve for flush valve water closets. With 324 SFU, the demand is about 97 GPM.
4. Read the solid-line curve for flush tank water closets. With 324 SFU, the demand is about 50 GPM.
5. Read the solid line curve for flush valve water closets. With 400 SFU, the demand is about 102 GPM.
6. Read the dashed curve for flush tank water closets. With 3,000 SFU, the demand is about 410 GPM.

4.3 SIZE SUPPLY

SIZE WATER SUPPLY PIPE

Gallons per minute (GPM) delivered by a main, branch, or riser depends on flow velocity in feet per second (fps). Faster flow means more GPM, more pressure loss due to friction (psi/100′), and more noise. Use figure 4.5 to find flow velocity (fps) and friction loss (psi/100′) for a 1/2″ pipe delivering 4 GPM. Notice that if flow is cut 50 percent (to 2 GPM), velocity drops from 4 to 2 fps, but friction loss drops from 7 to 2 fps.

A building's water supply piping can be sized with flow velocities as high as 10 feet per second (fps), but to minimize noise and pressure losses due to friction this text assumes a very conservative maximum flow velocity of *6 feet per second* for in-building piping and 8 fps for underground piping.

Look at Figure 4.5 again; when the required flow is 4 GPM, 1/2″ or 3/4″ pipe could be used. While 1/2″ pipe is cheaper, 3/4″ pipe will make less noise when water is flowing and generate less pressure drop due to pipe friction. Using a maximum flow velocity of 6 fps, 1/2″ pipe would be specified.

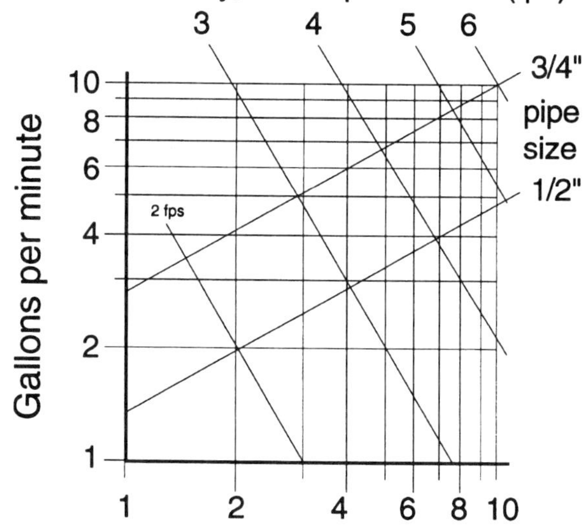

FIGURE 4.5 GPM and friction loss

EXAMPLES

Study Figure 4.5 for a moment and fill in the underlined values:

size	GPM	fps	psi/100′
1/2″	2	?	?
3/4″	10	?	?
3/4″	?	?	3

Check your answers on the following page after you use Figure 4.6 to answer the following questions.

1. Select a riser to deliver 310 GPM in a large hotel. Limit flow velocity to 6 feet per second and check the friction loss.
2. Select a water main to deliver 200 GPM to a 100-unit apartment. Limit flow velocity to 8 feet per second and check the friction loss.
3. Select a water main to deliver 100 GPM to a school. Limit flow velocity to 8 feet per second and check the friction loss.

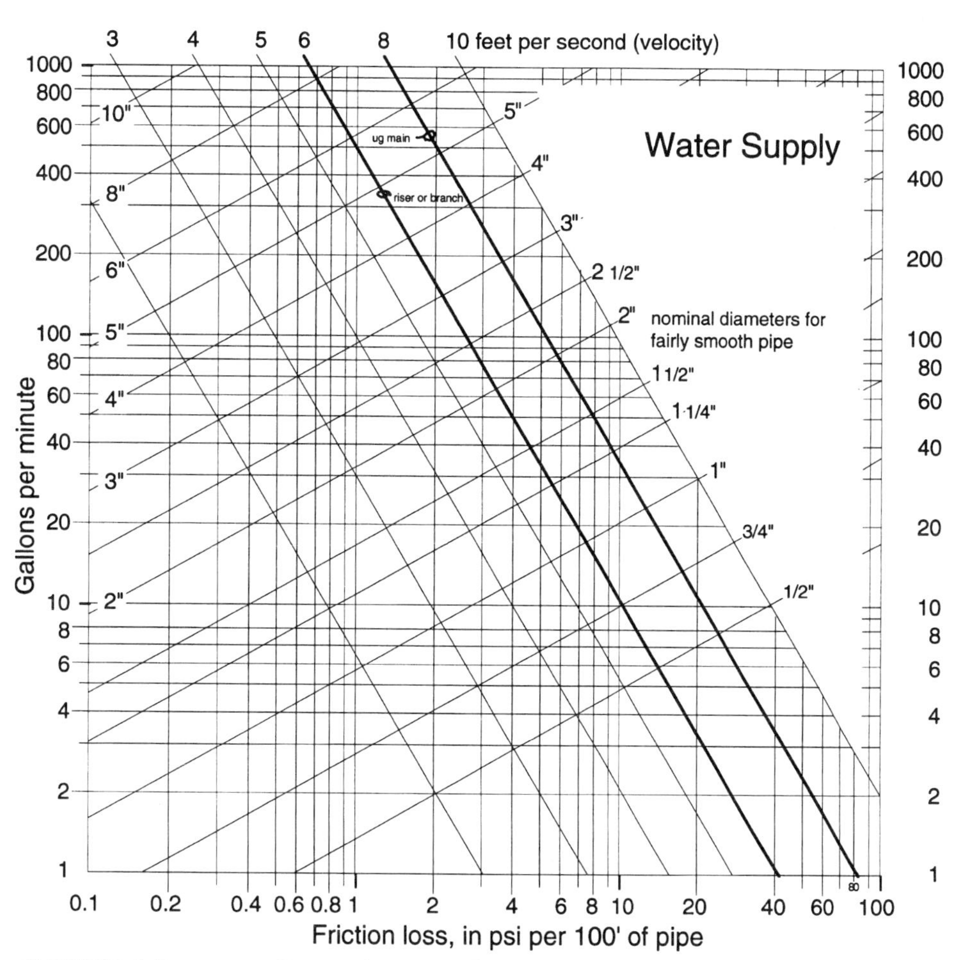

FIGURE 4.6 Piping flow, velocity, and pressure drop in fairly smooth copper or PVC pipe

4. Select a water main to deliver 50 GPM to a store. Limit flow velocity to 8 feet per second and check the friction loss.

ANSWERS

Size	GPM	fps	psi/100′
1/2″	2	2	2
3/4″	10	6	10
3/4″	5	3	3

1. Select 5″ pipe to deliver 310 GPM. Actual friction loss will be about 1 psi/100′. Many engineers would specify 4″ riser for economy even though flow will be noisy with a velocity of 8 fps.
2. Select 3″ pipe to deliver 200 GPM. Actual friction loss will be about 4 psi/100′, and velocity will be a little more than 8 fps.
3. Select 2½″ pipe to deliver 100 GPM. *Actual* friction loss will be about 3 psi/100′.
4. Select 1½″ pipe to deliver 50 GPM. Actual friction loss will be about 8 psi/100′.

SIZE RISERS AND BRANCHES

The building main is sized to meet expected demand for hot *and* cold water, but individual risers and branches are sized for hot *or* cold water demand. The actual friction loss (psi/100′) for the pipe size selected for the main is used to size *all* branches and risers.

Fixtures connected to both hot and cold water are rated at 75 percent for each—e.g., a lavatory rated at 4 SFU is counted as 3 SFU for cold water and 3 SFU for hot water.

EXAMPLE

A gymnasium main calculation established a friction loss of 5 psi/100′ for all pipe runs. Size hot and cold supply branches for a locker room with 1 flush valve WC, 1 Urinal, 2 Lav's, and 4 Showers.

Fixtures	*SFU*	*CW*	*HW*
1 WC @ 10	10	10	
1 Urinals @ 5	5	5	
2 Lav's @ 4	8	6	6
5 Showers @ 4	20	15	15
		SFU 36	SFU 21
Supply GPM required.		51	12

See Figure 4.4; read CW from valve curve—HW from tank curve
Branch sizes @ 5 psi/100′—Figure 4.6 *CW 2″, HW 1″*

4.4 METERS

Meters are the final component of the water supply system. Larger meters provide more GPM and larger meters add substantially to the system's first cost. Table 4.4 tabulates flow in GPM with meter pressure drops of 10 psi. and 4 psi. Where city pressure is adequate use the 10 psi. values, but use the 4 psi. values if supply pressure is less than 50 psi.

EXAMPLES

1. Select a meter to deliver 60 GPM if a 10-psi pressure drop is acceptable.
2. Select a meter to deliver 60 GPM if low supply pressure limits meter pressure drop to 4 psi.
3. A building with a flush valve system has a total of 350 SFU. Find the main size (Figures 4.4 and 4.6) if flow velocity is limited to 6 fps, and find the meter size (Table 4.4) if a 10-psi drop through the meter is acceptable.

ANSWERS

1. 1½″
2. 2″
3. Select a 2½″ main (from Figure 4.4 GPM is about 95, from Figure 4.6, @ 95 GPM select 2½″ @ 6fps)
   Select a 2″ meter
   (yes, you can use a main larger than the meter to limit friction losses, velocity, and noise)

TABLE 4.4

Meter GPM and Pressure

Meter size	*10 psi loss GPM*	*4 psi loss GPM*
5/8″	12	8
3/4″	21	14
1″	33	20
1½″	63	40
2″	100	63
3″	200	125
4″	350	200
6″	625	440

4.5 SIZE DWV

DWV pipe size increases as the number of dfu carried increases. Review Tables 4.5 through 4.8 and then use them to complete the sizing examples.

TABLE 4.5

DWV Minimums

Drains

drainpipe size	1¼″
branch size below the floor	2″
allow 1/4″ per foot fall for waste and soil branches	
water closet outlet	4″
a 3″ drain can serve 2 WC's, but 4″ preferred	

Vents

vent pipe size	1¼″
individual vent	1/2 size of trap served
circuit vent	1/2 size of drain branch served
vents > 40′ long	increase to next std. pipe size
vents > 100′ long	increase another std. pipe size
std. pipe sizes 1¼″ - 1½″ - 2″ - 2½″ - 3″ - 4″ - 5″ - 6″ - 8″ - 10″	

TABLE 4.6

Building Drains

Maximum DFU for *Building Drains* condensed from the National Plumbing Code

Pipe size	*Fall in inches per foot of run* 1/16″	1/8″	1/4″	1/2″
2″			21	26
2½″			24	31
3″		36*	42*	50*
4″		180	216	250
6″		700	840	1,000
8″	1,400	1,600	1,920	2,300
10″	2,500	2,900	3,500	4,200

*not more than 2 water closets

TABLE 4.7

Soil and Waste Branches and Stacks

Maximum DFU for *Branches*** & *Stacks* condensed from the National Plumbing Code
BI = Branch Interval (each building story with drain branches is counted as one BI)

Pipe size	*Branch*	*Stack-3BI or less*	*Stack > 3BI total*	*each BI*
2″	6	10	24	6
3″	20*	48*	72*	20*
4″	160	240	500	90
6″	620	960	1900	350
8″	1400	2200	3600	600
10″	2500	3800	5600	1000

*Not more than 2 water closets
**Except branches of the building drain

TABLE 4.8

Vent Stacks

Maximum *Vent* length* - set by soil stack DFU condensed from the National Plumbing Code

Soil-Waste Stack (dfu)		*Vent Pipe Size* 3″	4″	6″	8″
3″	(60)	400′			
4″	(500)	180′	600′		
6″	(1900)	20′	70′	700′	
8″	(3600)		25′	250′	800′
10″	(5600)			60′	250′

*Field numbers are the maximum vent length in feet

EXAMPLES

1. A 5-story building has a total of 400 DFU (80 per floor). Size the building drain, soil stack, and vent stack if the building is 80′ tall.
2. A 4-story building has a total of 400 DFU (100 per floor). Size the building drain, soil stack, and vent stack if the building is 60′ tall.
3. Size a horizontal branch to carry 4 lavatories rated at 2 DFU each.

4. Size a building drain to carry 4 lavatories rated at 2 DFU each.
5. Size a horizontal branch to carry 3 water closets rated at 3 DFU each.

ANSWERS

1. Building drain = 6″ @ 1/8″ per foot fall, soil stack = 4″, vent stack = 3″.
2. Building Drain = 6″ @ 1/8″ per foot fall, Soil Stack = 6″, Vent Stack = 4″.
   (soil stack is 6″ because a 4″ stack will carry a maximum of 90 DFU per floor.).
3. 3″ branch
4. 2″ building drain @ 1/4″ fall if all lavatories are connected to the building drain, but 3′ if lavatory branch is not part of the building drain.
5. 4″ branch (a 3″ branch will carry 20 DFU, but not more than 2 water closets)

4.6 EXAMPLE RESIDENCE

The example residence in Figure 4.7 is a 1,568-sq. ft. slab on grade structure used in companion texts to illustrate lighting, electrical, and HVAC work. Plumbing work for the house will consume 8 to 10 percent of the construction budget if municipal water and sewer are available. A municipal plumbing permit costs hundreds of dollars and usually includes inspections and connections to city water and sewer lines. On building sites outside the city limits, a well and septic system will add thousands of dollars to the project budget.

FIXTURE LOCATION

Codes require a minimum of 1 bathroom and 1 kitchen sink in each residence. The example will include plumbing fixtures and connections for two bathrooms, kitchen, laundry, and exterior hose bibs (Figure 4.8). Where handicapped access is required, designers consult Uniform Federal Accessibility Standards for dimensions, clearances, and grab bar requirements. Experienced designers locate fixtures on interior partition walls so that DWV trees don't interfere with foundations, structure, and fascia, but the example house has a window above the kitchen sink so wall framing there must be reinforced.

PLUMBING PLAN

Complete plumbing plans and isometrics are typical in commercial construction, but for many residential projects the plumber develops a piping layout working

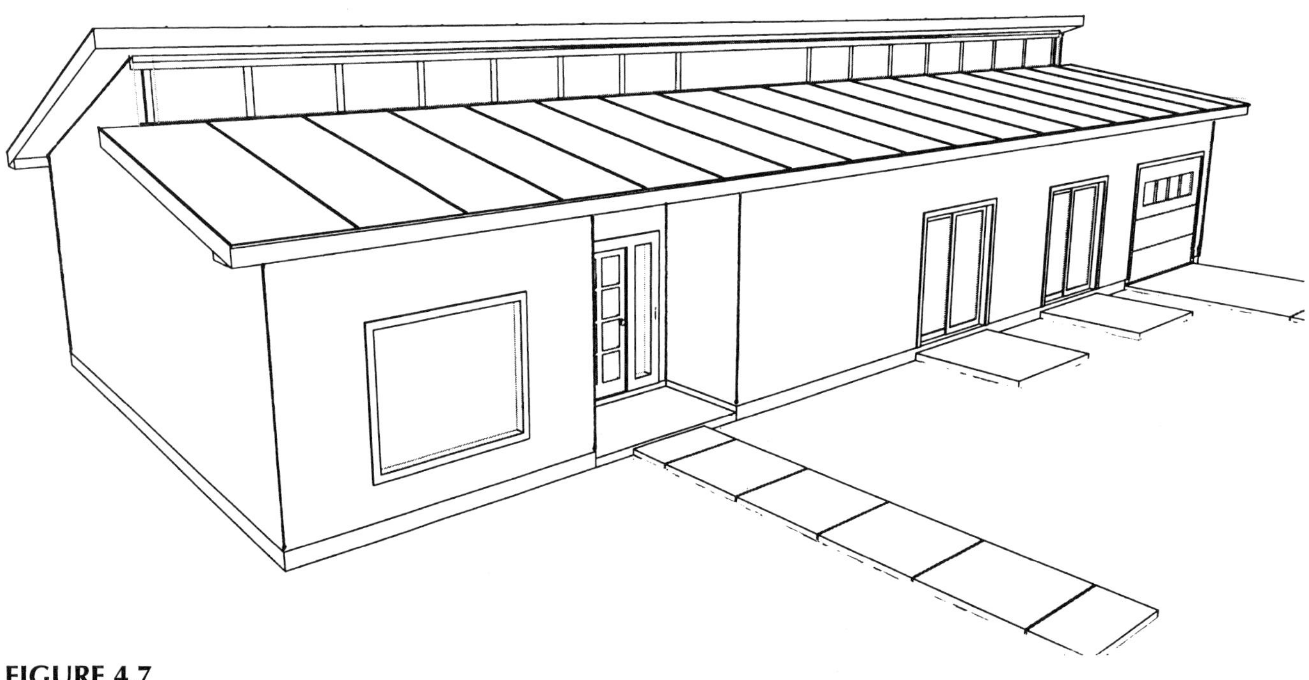

FIGURE 4.7

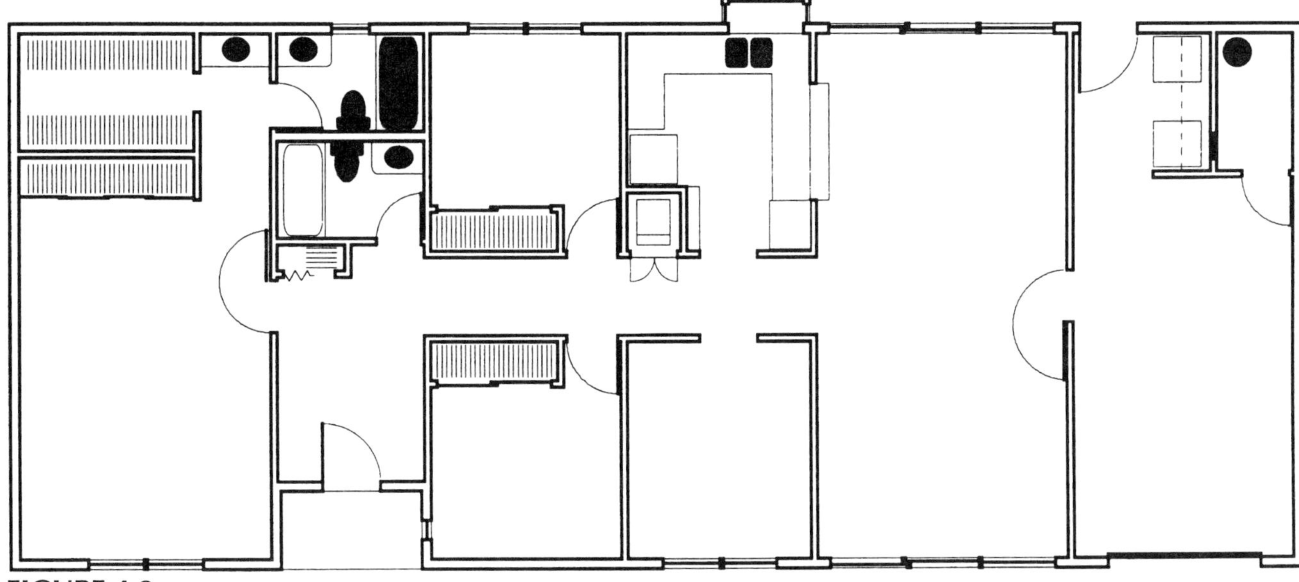

FIGURE 4.8

from the fixture locations shown on the architectural plans. Plumbers consult the city inspector to find water and sewer lines that will serve the home, and then locate site trenches offering the shortest route to avoid existing tree roots and driveways (Figure 4.9).

UNDERGROUND

Fixture waste and soil connection points are carefully set to meet the architectural plans. Waste and soil pipe runs fall 1/4″ per foot as they dodge foundations, grade beams, and piers (Figure 4.10). Floor drains must be set to elevation and anchored before placing concrete (2″ floor drains must be vented within 6 feet of the trap).

All joints below grade are leak tested before backfilling. Underground hot and cold water supply piping is often run with soft temper copper loops because no fittings are required below the floor. Supply loops rise along the center lines of future wall framing, where the plumber sleeves, seals, and identifies each loop.

ROUGH IN

The plumbing crew begins "rough in" when the slab, roof deck, and exterior sheathing are complete. Water supply and drain pipes are stubbed at the proper height for individual fixtures, and DWV trees are completed through the roof. A 3″ VTR (vent through roof) located in a 2×6 stud partition wall serves the bathroom DWV tree; vent flashing is usually provided by the plumber. Smaller VTR trees serve the kitchen and laundry (Figure 4.11).

Two bathtubs are set in alcoves faced with water-resistant sheetrock. Access panels for tub-shower valves are located in the entry closet and behind the medicine cabinet and below the lavatory in the guest bath.

A laundry supply-waste box is installed in the wall behind the washer, and water lines are run up the laundry wall to serve the washer and the water heater. Pipe stub-outs at each fixture are securely anchored to permit installation of threaded fittings for fixture service valves.

Black steel natural gas outlets are stubbed at the water heater and the furnace and all piping is leak tested again, before insulation and sheetrock work begins.

SET AND FINISH

When interior tile, walls, ceilings, and cabinets are complete, the plumbing crew returns to set toilets, sinks, hose bibs, and the water heater. Valves, trim, and the garbage disposer are installed and a final inspection is scheduled.

City personnel usually make or supervise connections to city water and sewer mains, supply piping is flushed with chlorinated water, and all fixtures are checked for proper operation.

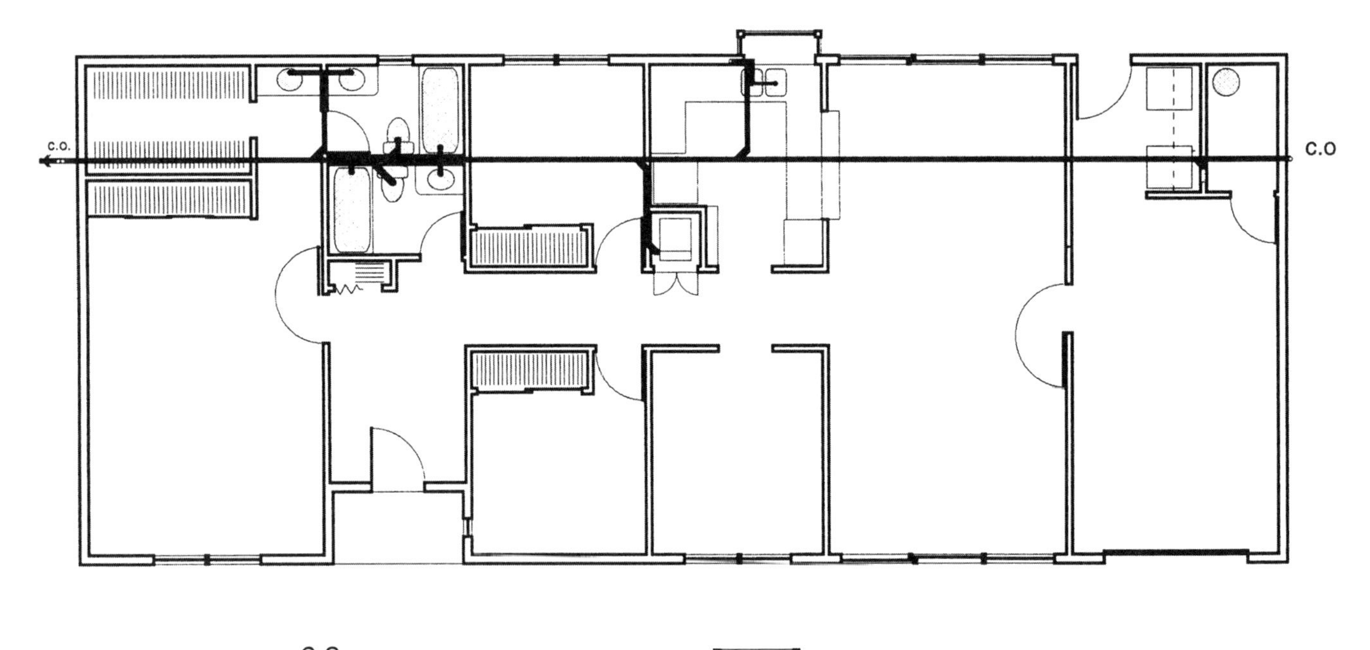

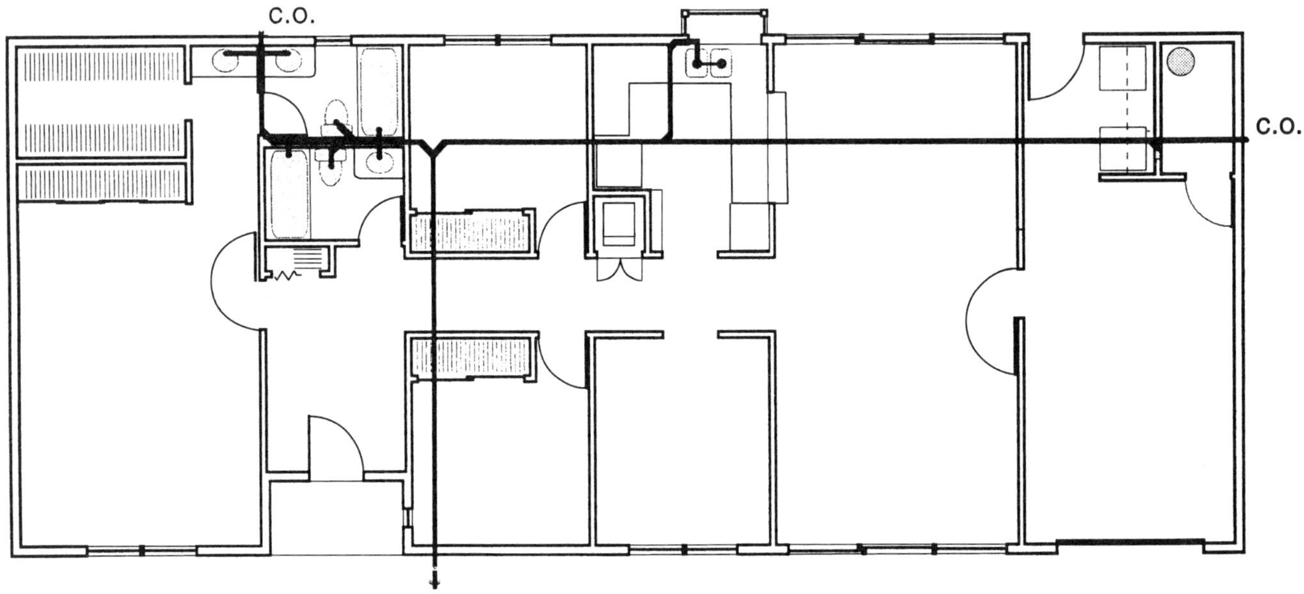

FIGURE 4.9

CHECKLIST

The following items are usually covered in the plumbing specifications.

- Condensate drain for AC unit(s)
- Drain pan under AC (if in attic)
- Drain pan under water heater (in attic)
- Water heater pressure relief valve (check local codes—some codes require PRV discharge outdoors, other codes allow discharge via drain line)
- Water hammer (consider eliminators at each fixture for flush valve systems and locations with high supply pressure)
- Pressure reducing valve (consider in locations where city pressure is high; supply landscape outlets at full pressure and locate the pressure reducing valve to supply the indoor fixtures)
- Freeze protection (drain valves or special fittings) for hose bibs or fixtures in unheated locations

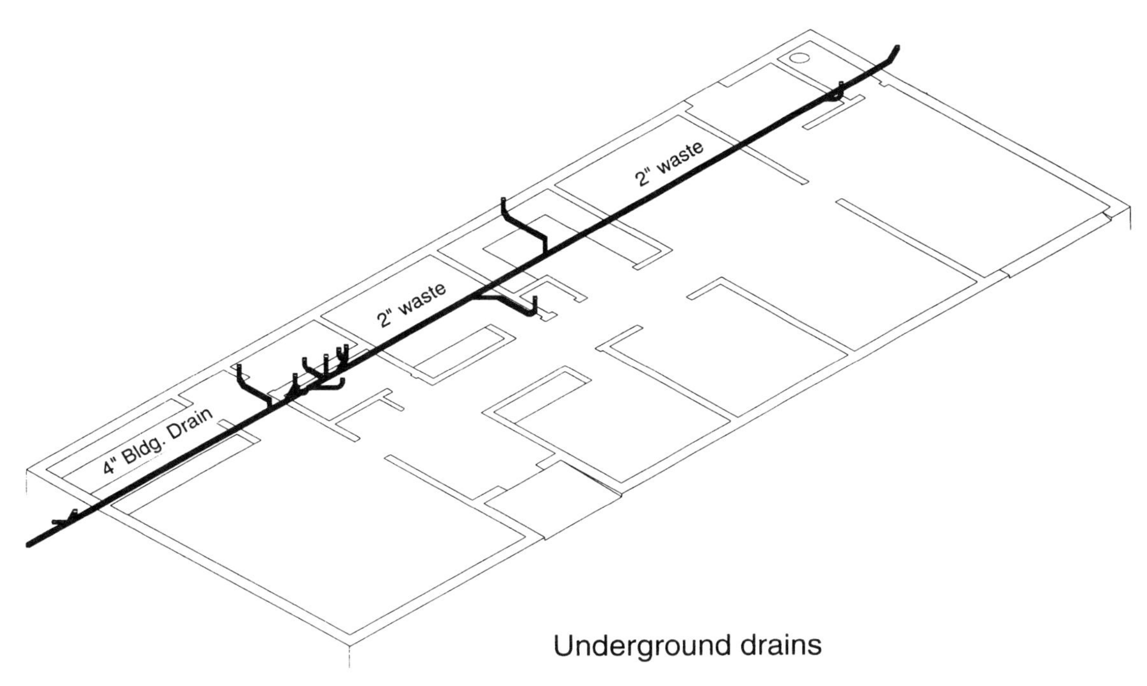

FIGURE 4.10 Underground drains

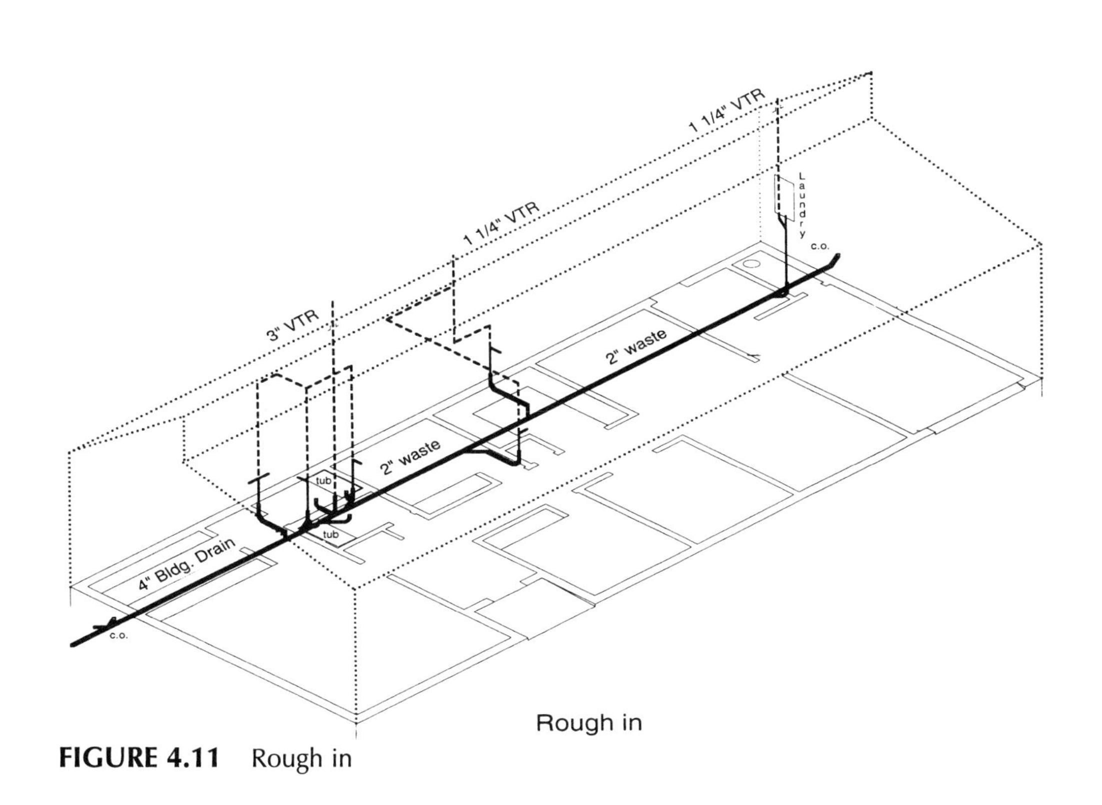

FIGURE 4.11 Rough in

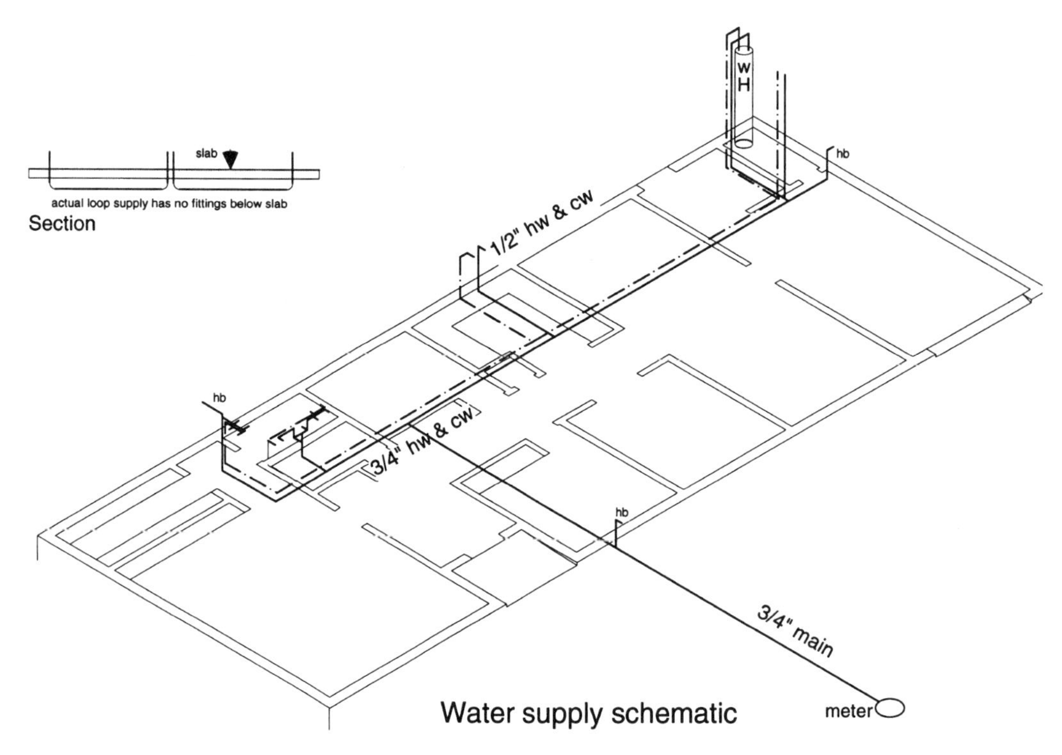

FIGURE 4.12 Water supply schematic

SIZE EXAMPLE RESIDENCE PLUMBING

Sizing residential plumbing is not difficult. Most homes are served by a 3/4″ meter* so supply piping choices are 1/2″ or 3/4″. DWV size choices range from 1¼″ to 4″. Most local codes require a 4″ house sewer, and branch drains get smaller when they carry fewer DFU. Most vents are 1¼″ except for a code-required 3″ VTR that usually serves the toilets.

Water Supply The 3/4″ water main should run full size to hose bibs, the water heater, and any group of fixtures that demands more than 4 SFU (supply fixture units). Smaller 1/2″ lines can serve lavatories, kitchen sinks, and a tub-shower or a toilet at the end of a run. 3/8″ pipe is NOT used for "rough in" but flexible 3/8″ lines are used to connect fixtures to their service valves (Figure 4.12).

The example house uses pressure tank water closets, so the pressure at these toilets must be checked after the main friction and head losses are established (explained further in the following subsection headed Pressure Check).

Waste and Soil In a single-story residence all waste and soil lines are part of the building drain. A 2″ drain will carry 20 DFU (drain fixture units); 3″ will carry 30 DFU but not more than 2 WC's; and a 4″ drain can carry 160 DFU. Follow the house sewer line from right to left starting at the 2″ laundry inlet. A 2″ drain branch can carry 20 DFU, so it easily carries the laundry, kitchen sink, tub-shower, and lavatory (8 DFU). Increase it to 4″ at the first toilet connection since most local codes require a 4″ sewer beyond the building line. A 3″ drain could serve this house but the cost difference between 15′ of 3″ pipe and 4″ pipe is trivial and 4″ is less likely to be clogged (Figure 4.13).

*5/8″ meters are used by some water supply districts and 1″ meters are purchased by affluent homeowners with swimming pools and lawn sprinkler systems.

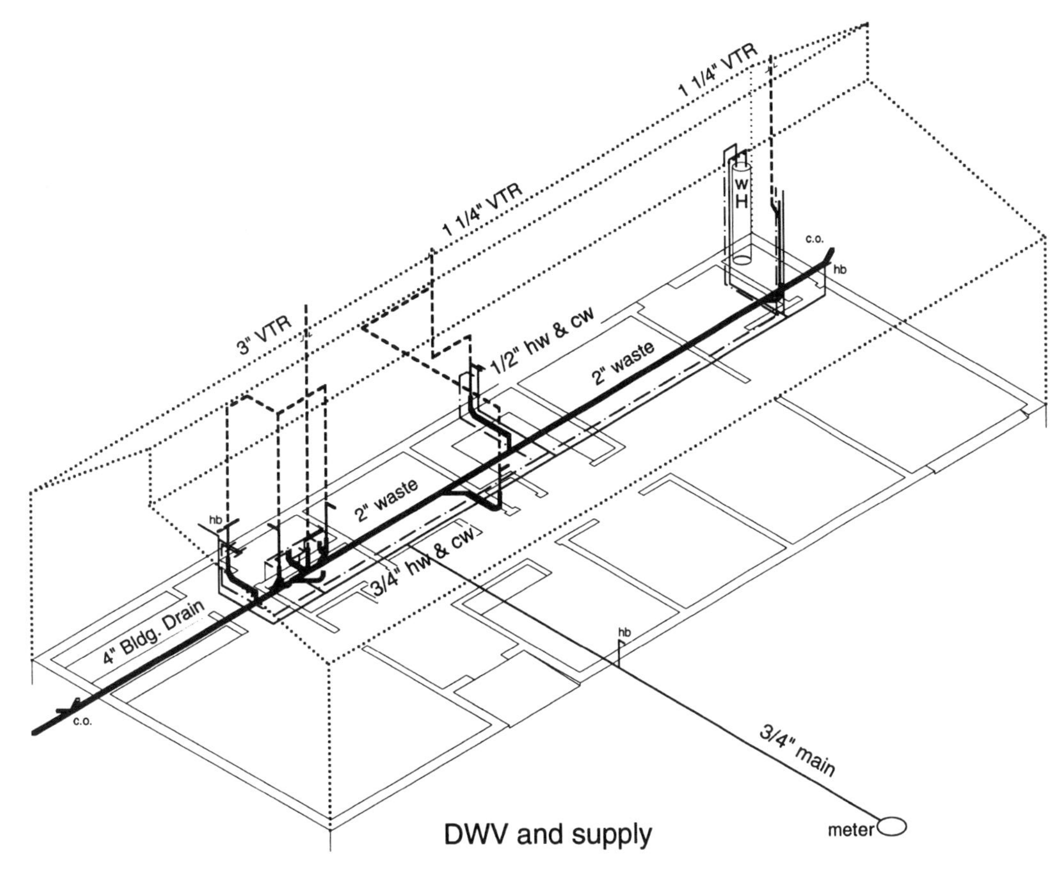

FIGURE 4.13 DWV and supply

Vents Individual vents must be 1/2 the size of the trap they serve, but not less than 1¼″. Code requires one 3″ VTR (vent through roof) that's usually connected at the toilet outlets. Surrounding fixture vents join the 3″ VTR in the wall or attic to minimize roof penetrations.

Fixture Connections	*supply*	*drain*	*vent*
Lavatory	1/2″	1¼″	1¼″
Laundry box	1/2″	2″	1¼″
Tub - Shower	1/2″	1½″	1¼″
Water Closet(s)	1/2″	4″	3″*

*One 3″ vent required by code

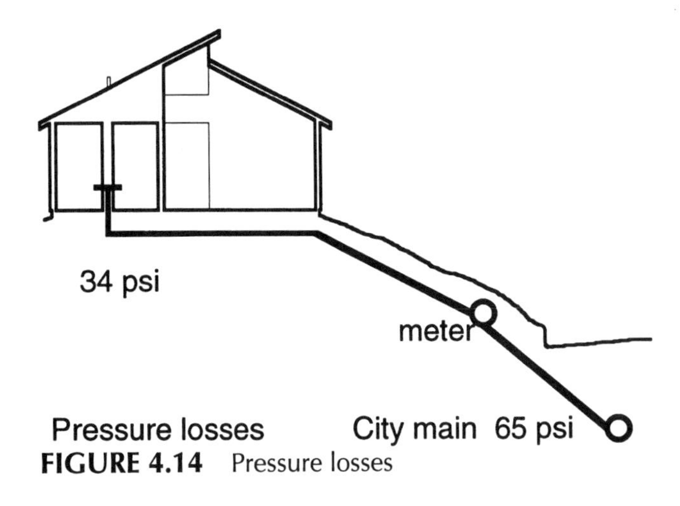

FIGURE 4.14 Pressure losses

PRESSURE CHECK

Pressure tank toilets specified for the example residence need 25 psi. to flush effectively. Pressure losses between the Municipal water supply and the toilet inlet must be tabulated to verify adequate residual pressure (Figure 4.14).

Given: The pipe run from city main to most distant toilet measures 80′, but allow 50 percent extra for friction losses in the pipe fittings. Total friction losses will be calculated for a 120′ run.
The City main flows @ 65 psi, and its elevation is 9′ below the pressure tank toilet inlets.

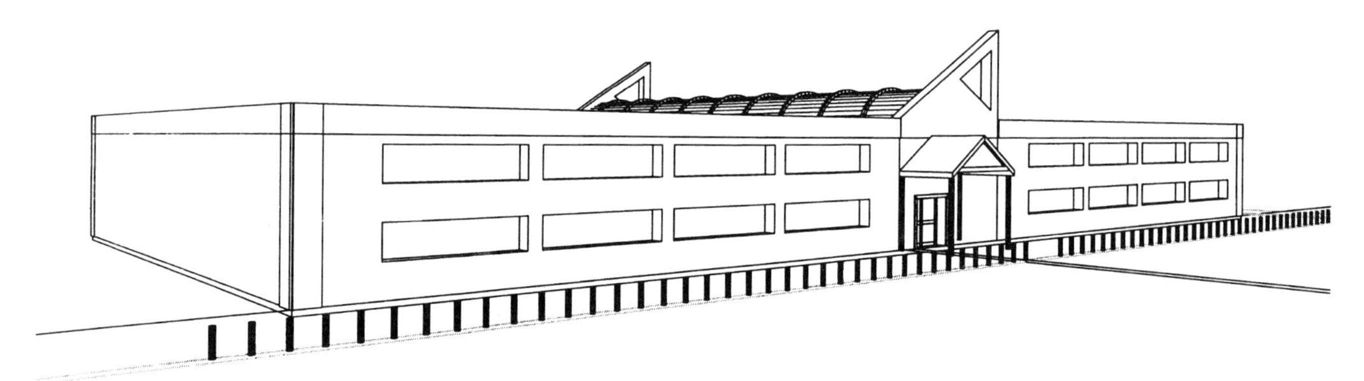

FIGURE 4.15

From sections 4.3 and 4.6:

Maximum home demand — about 14 gpm
(15 SFU excluding hose bibs* — table 4.4)
Friction pressure loss — about 19 psi/100′
(3/4″ pipe,19 psi/100′ @ 14 GPM — Table 4.6)

Calculate pressure losses:

head (9)(0.434) = 3.9, say 4 — 4 psi
(meter elevation is 9′ below toilet tank)
meter — 4 psi
(3/4″ meter @ 14 GPM — allow 4 psi; see Table 4.7).
pipe friction (19 psi/100′)(120′) = 22.8 — 23 psi
(3/4″ pipe @ 14 GPM — see Figure 4.6)
Total pressure loss 31 psi

Required toilet operating pressure 25 psi
Pressure at toilet inlets (65 − 31) 34 psi

The available pressure is adequate. If the pressure at the toilet inlet was less than 25 psi, a booster pump would be required.

4.7 EXAMPLE OFFICE BUILDING

The example office is a 21,600-sq. ft. two-story structure used in Volumes 1 & 2 to illustrate lighting, electrical and HVAC installations (Figure 4.15). Where city water and sewer connections are available, plumbing costs for this office will consume 4 to 6 percent of the construction budget.

Most building plumbing is concentrated in the 4 rest rooms adjacent to the two story entry lobby (Figure 4.16). Men's and women's rest rooms on both floors are separated by the main circulation corridors, and the kitchens and drinking fountains (not shown) are located near individual office entries. Pipe sizing calculations that follow include allowances for all building plumbing fixtures, but drawings detail only the rest room area.

FIXTURES REQUIRED

Table 4.1 shows 1 occupant per 150 sq. ft., or 144 people, but allow for a total building occupancy of 160 as estimated in previous HVAC calculations.

Minimum fixtures required to serve the building population are 5 WCs and 3 Lav's for women, 3 WC's, 2 UR's, and 3 Lav's for men, and 2 DF's. These fixture requirements are developed from Table 4.2, allowing for 96 women and 96 men (60 percent of 160).

The building owner orders more fixtures—12 WC's , 8 Lav's, 4 UR's, 4 Kitchen Sinks, 2 Service Sinks, and 4 DF's—to accommodate occupants of a planned future addition (Table 4.9).

Plumbing the office requires three separate labor activities: underground work, rough in, and set and finish. These activities were described in detail for the example house on preceding pages, so only differences for the office are covered.

*House fixtures total 15 SFU plus 16 SFU for hose bibs. If you calculated the house main to carry 31 SFU (17 GPM), you would select a 1″ main, and pay more for a 1″ meter (or do a challenging calculation for flow through a 3/4″ meter with a 1″ main, or schedule hose bib use when your showers and laundry are inactive).

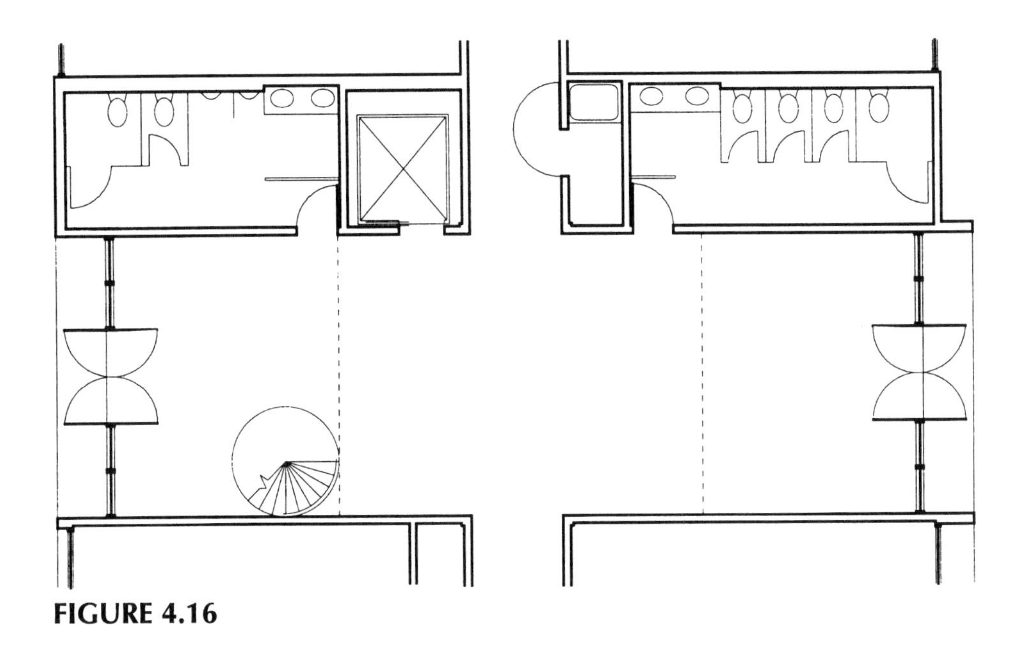

FIGURE 4.16

TABLE 4.9

SFU and DFU from Table 4.3 (fixture requirement)

Fixture	*SFU*	*DFU*	*psi*
WC's, flush valve (12)	120	72	25
Lav's (8)	16	8	10
UR's, flush valve (4)	20	16	15
DF's (4)	1	2	10
Kitchen Sinks (4)	16	12	10
Service Sinks (2)	6	6	10
Hose Bibs (6)	*	—	10+
Landscape water	*	—	10+
	totals	179	116

*SFU are not counted for hose bibs or the building's landscape watering system because their operation will be scheduled when fixture water demand is low.

Underground Underground DWV piping is similar to the example house whether the first floor is slab on grade construction or spans over a crawl space. If a full basement is used, a sump pump and a sewage ejector pump may be required depending on the elevation of the city sewer.

The water supply main will enter the building below the rest room plumbing walls. Two risers, sized for quiet flow, will serve the rest rooms unless the building is sprinklered. Mains for a sprinklered building are sized for fire protection; they enter the building at a location where firefighters can access control valves, and the smaller domestic water main is connected there.

Rough In DWV trees repeat on each floor, so separate soil and vent stacks are required. The vent stack connects to the soil stack above the highest vent branch, and in high-rise buildings relief vents connect vent branches and soil branches on all except the top floor.

Supply branches run horizontally in the rest room walls, and piping that serves remote fixtures is usually run above the ceiling. Natural gas piping in multistory commercial buildings is run in fire-resistant vertical chases with natural ventilation openings above the roof.

Set and Finish When interior walls, ceilings and cabinets are complete, the plumbing crew returns to set toilets, sinks, water heater, etc., and to install the valves and trim.

Supply piping is flushed with chlorinated water, all fixtures are checked for proper operation, and after final inspection, connections to the city water and sewer mains are completed.

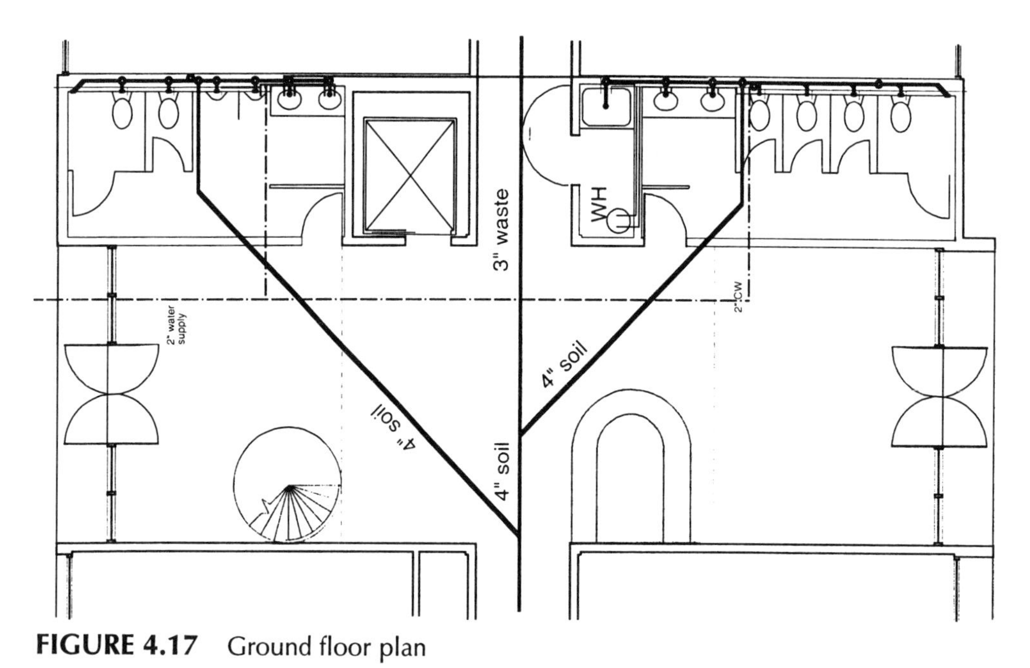

FIGURE 4.17 Ground floor plan

PLUMBING PLAN AND ISOMETRIC

The plan and isometric shown in Figures 4.17 and 4.18 define the plumbing work in the example office rest rooms. The plan illustrates ground floor piping and the isometric details pipe sizes. Pipe size calculations can be found on the following pages.

DWV The building drain runs from one end of the building to the other. It carries waste from the kitchens and drinking fountains (not shown) and picks up the rest room drains below the lobby floor. Building drain DFU's and sizes are shown on the plan; stack and branch piping sizes are on the isometric. Circuit vents are shown in the women's rooms and individual vents are illustrated in the men's rooms. Some local codes prohibit circuit venting.

Supply The building supply main splits below the lobby floor to serve the men's and women's rest rooms. The women's room supply riser carries all building hot water and both service sinks. Water lines for kitchens, DF's and remote hose bibs are run overhead.

DWV PIPE SIZES

A 4″ drain sloped 1/8″ per foot will be used to carry the building's 116 DFU (Table 4.6). 4″ branches and soil stacks will also serve each rest room because 3″ pipe cannot carry more than 2 WC's (Table 4.7). A 3″ branch must be used to carry 2 kitchen sinks and 2 DF's (7DFU). 3″ vent stacks are required to serve each restroom soil stack (Table 4.8).

The remaining DWV piping is sized using Table 4.8.

SUPPLY PIPE SIZES

GPM Figure 4.4 shows a demand of about 80 GPM for the building's 179 SFU.

Main and Meter Size Use Table 4.6 to select a 2″ main with a friction loss of 5 psi/100′. Flow velocity in the main is just under 8 fps, and by splitting the main as it enters the building the branches and risers can be sized at a quiet 6 fps.

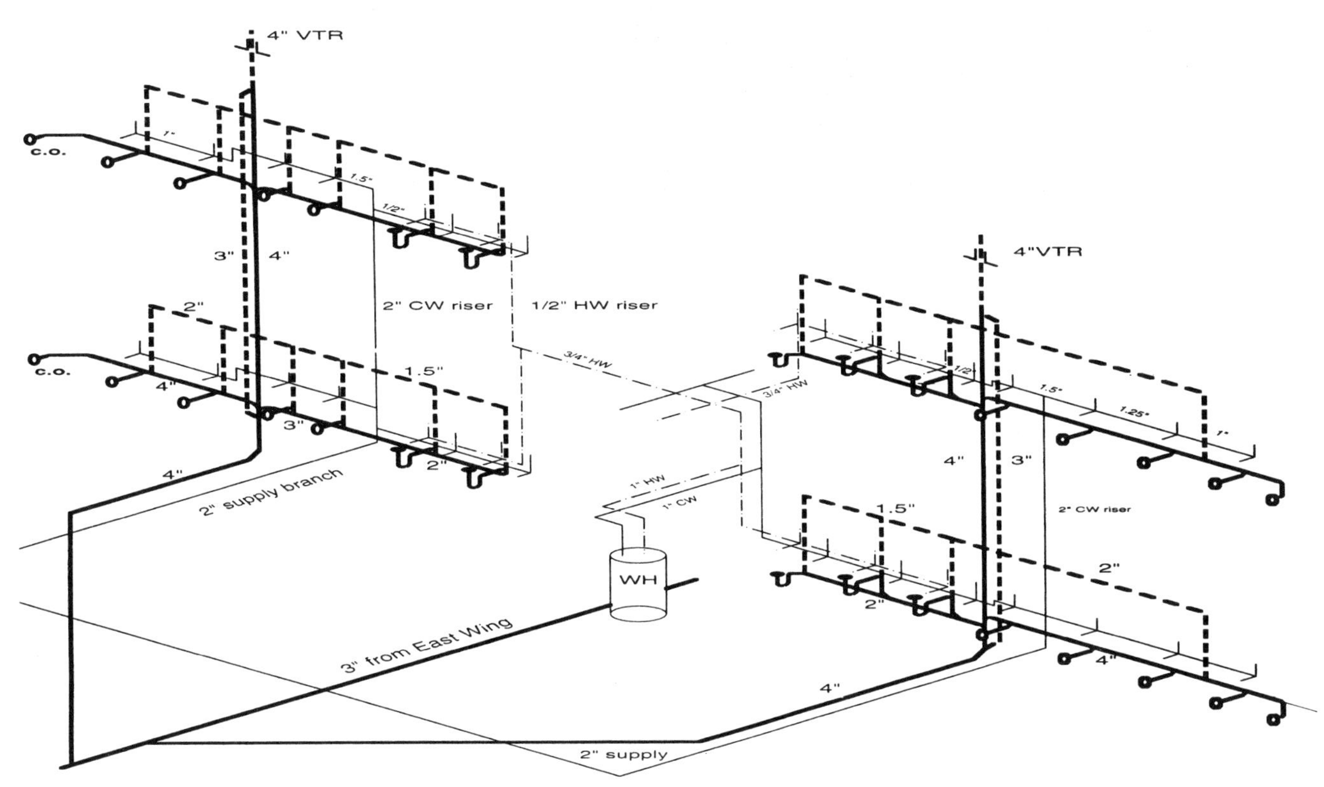

FIGURE 4.18 Isometric plan with pipe sizes
Women's rooms below right (circuit vents), men's rooms below left (individual vents).
DWV soil and waste = heavy solid lines, vents = heavy dashed lines. Sizes are bold numbers.
Cold water = light solid lines, hot water = light dashed lines. Sizes are light numbers.
Vertical lines at each fixture are water hammer eliminators.

Check Operating Pressure The most demanding fixture in the building is the second-floor flush valve WC that is most distant from the city main. If there is a residual pressure of 25 psi at this WC, pressure will be more than adequate at all other fixtures.

Given: The pipe run from City main to the most distant second floor WC is 200′, but allow 50 percent extra for friction losses in the pipe fittings. Total friction losses will be calculated using a developed pipe length of 300′.

The City main flows at 60 psi. Its elevation is 20′ below the flush valve WC's on the second floor.

From Tables 4.3 and 4.6:

Maximum demand	80 GPM
(179 SFU — Table 4.3)	
Pipe friction pressure loss	5 psi/100′
(2″ pipe, 5 psi/100′ @ 80 GPM — Table 4.6)	

Calculate pressure losses:

head (20)(0.434) = 8.7, say	9 psi
(main elevation is 20′ below toilet tank)	
meter	7 psi
(2″ meter @ 80 GPM — interpolate — Table 4.7).	
pipe friction (5 psi/100′)(300′) = 15	15 psi
total pressure loss 31 psi.	
Required toilet operating pressure	25 psi
Pressure at toilet inlets (60 − 31)	29 psi

The available pressure is adequate. If the pressure at the toilet inlet was less than 25 psi a larger main (with less friction loss) or a booster pump would be required.

SIZE RISERS AND BRANCHES

Risers and branches are sized with the same 5 psi/100′ friction loss calculated for the main. This assures adequate pressure at each fixture and allows branch and riser size to be reduced as each supplies fewer fixtures. Pipe sizes for in-building risers and branches are also checked for flow velocity; if velocity exceeds 6 fps, the next larger pipe size is selected.

The 2″ building main supplies up to 80 GPM (179 SFU). It connects to the landscape water system outside the building and then divides into two risers below the ground floor. One riser serves the women's rest rooms on two floors, the water heater, and the service sinks; the second riser serves the men's rest rooms, DF's, hose bibs, and the kitchen sinks.

Riser One — 112 SFU @ ground floor
8 WC (80SFU), 4 Lav's (8 SFU), 2 Service Sinks (6 SFU), plus hot water* for 4 Lav's (6* SFU) and 4 Kitchen Sinks (12* SFU).

Riser One — 53 SFU @ second floor
4 WC (40 SFU), 2 Lav's (4 SFU), 1 Service Sink. (3 SFU), and 2 Kitchen Sinks (6 SFU).

Riser Two — 78 SFU @ ground floor
4 WC (40 SFU), 4 UR (20 SFU), plus cold water* for 4 Lav's (6* SFU), and 4 Kitchen Sinks (12* SFU).

Riser Two — 39 SFU @ second floor
2 WC (20 SFU), 2 UR (10 SFU), plus cold water* for 2 Lav's (3* SFU), and 2 Kitchen Sinks (6* SFU).

Use Table 4.4 to convert SFU to GPM, then size risers at 5 psi/100′ using Table 4.6.

Riser	*SFU*	*GPM*	*Pipe*
One—ground floor	112	70	2½″
One—2nd floor	58	60	2″
Two—ground floor	78	65	2″
Two—2nd floor	39	55	2″

Wow! All that work and no smaller pipe, but what about noise? Redo these risers, limiting maximum velocity to 6 fps.

*HW & CW SFU = 3/4 of total fixture SFU

Use Figure 4.6 and revise riser size to limit flow noise. Size risers at 6 feet per second maximum velocity.

Riser	*SFU*	*GPM*	*Pipe*
One—ground floor	112	70	2½″
One—2nd floor	58	60	2″
Two—ground floor	78	65	2″
Two—2nd floor	39	55	2″

At the end of supply runs, where a riser or branch serves just a few fixtures, you cannot read GPM on the demand curves. In these cases size supply piping as follows:

One flush valve WC — minimum 1″ — two or three flush valve WC's, use 1¼″ pipe.

One flush valve Urinal — minimum 3/4″ — two to four, use 1″ pipe.

Lavatories, Showers, Sinks, etc., use 1/2″ up to 4 SFU and 3/4″ up to 12 SFU.

4.8 TALL BUILDINGS

Increasing building height increases head losses; above 4 or 5 stories (depending on the available supply pressure), water storage tanks and/or pumps will be components of the domestic water supply system. Downfeed systems pump water to a roof tank and rely on gravity distribution. Upfeed systems maintain pressure with pumps and are typical in recent high-rise buildings (Figure 4.19).

Codes require standpipes and sprinklers in most high-rise buildings (more than 75 feet tall). Standpipe systems are designed to deliver lots of water to fire hose cabinets located on each floor. The domestic water system is metered and previous pipe size calculations are valid, but a large separate main is required to serve standpipes (Figure 4.20).

High-rise buildings store water to fight a fire for 30 minutes, while firefighters make pumper and hydrant connections. A standpipe must deliver 500 GPM at 65 psi minimum for 30 minutes. For an installation with multiple standpipes, allow for 1,250 GPM, which is 150 tons of water over 30 minutes, so large storage tanks are required (Figure 4.21).

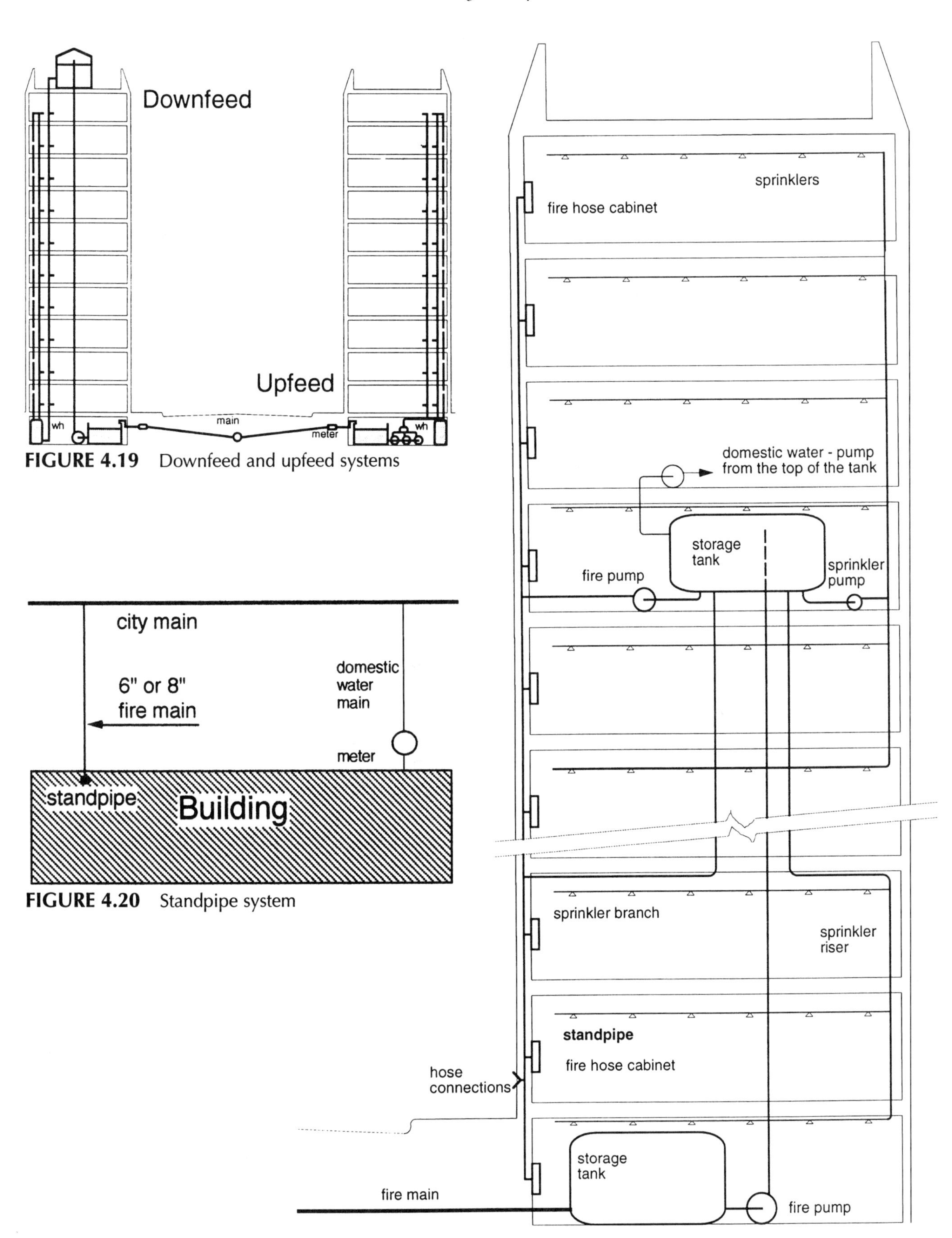

FIGURE 4.19 Downfeed and upfeed systems

FIGURE 4.20 Standpipe system

FIGURE 4.21 High-rise building with storage tanks

REVIEW PROBLEMS

1. Find the friction loss in a 200′ run of 2½″ pipe delivering 80 GPM.
2. How many GPM will a 1½″ pipe deliver with a friction loss of 10 psi/100′?
3. How many GPM will a 1½″ pipe deliver with a friction loss of 1 psi/100′?
4. A 2″ pipe must deliver 30 GPM. What is the friction loss over a 600′ pipe run?
5. A 2″ pipe must deliver 30 GPM. Find the water velocity in feet per second.
6. A 700′ length of 1½″ pipe is delivering 40 GPM. Find the water velocity in feet per second and the *total* pressure loss due to friction.
7. If water velocity is limited to 6 feet per second, find the maximum GPM that can be delivered by 1″, 2″, and 8″ pipe.
8. If friction loss is limited to 1 psi/100″ feet, find the maximum GPM that can be delivered by 1″, 3″, and 6″ pipe.
9. Can a 1″ pipe deliver 50 GPM? Why is a 2″ pipe a better choice at 50 GPM?
10. A service sink has a 3″ trap. Size its vent.
11. Size a building drain to carry 800 DFU with a slope of 1/4″ per foot.
12. Size a soil branch to carry 600 DFU.
13. Size a soil stack to carry 400 DFU.
14. A 6-story soil stack carries 400 DFU. What is the maximum DFU at any floor?
15. A soil stack carries 400 DFU. Find the vent stack size if the stacks are 150′ tall
16. A soil stack carries 400 DFU. Find the vent stack size if the stacks are 200′ tall.
17. A 2″ vent branch will be 50′ long. Should the vent size be increased?

ANSWERS

1. 4 psi — (2 psi/100′)(200′) = 4
2. 60 GPM
3. about 11 GPM
4. about 6 psi
5. 3 fps
6. velocity = 6 fps; the total pressure loss is about 31 psi.
7. 1″, = 16 GPM; 2″, = 64 GPM; 8″, = 900 GPM
8. 1″, = 5+ GPM; 3″, = 90 GPM; 6″, = 550 GPM
9. Yes a 1″ pipe can deliver 50 GPM, *but* water velocity will be more than 10 feet per second and friction loss will be about 55 psi per 100′. 2″ pipe is a much better choice because flow velocity is less than 5 fps and friction losses are reduced by 95 percent.
10. 1½″
11. 6″
12. 6″
13. 6″ for 3 stories or less, 4″ if more than 3 stories.
14. 90
15. 3″
16. 4″
17. Yes, increase to 2½″

Water Supply for Fire Protection and HVAC

The steaming river has washed the hot round red sun down under the sea.

Basho

This chapter deals with water supply systems for two other important uses of water in buildings—fire protection, and heating and cooling. Water is a very economical and readily available medium for suppression of fire. The capability of water to carry large quantities of heat also makes it a popular fluid for use in HVAC, Heating, Ventilating, and Air Conditioning systems.

5.0 WATER FOR FIRE PROTECTION

BASIC FACTORS OF FIRE

Fire or combustion is a chemical reaction involving fuel, oxygen, and high temperature (Figure 5.1). In this reaction, molecules of a fuel are combined with the molecules of oxygen in a reaction that also results in an evolution of light and energy. If any of the three factors involved in the reaction is removed, a fire will be extinguished.

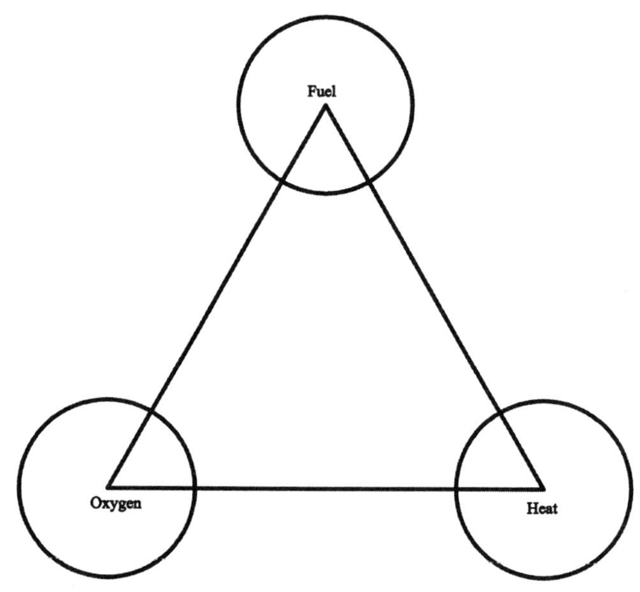

FIGURE 5.1 The fire triangle

CLASSIFICATION OF FIRE HAZARDS

The National Fire Protection Association (NFPA) has categorized fire hazards into four different classes: A, B, C, and D. Class A fires involve solid combustibles (such as wood or paper), which are best extinguished by water or dry chemicals. Fires in flammable liquids are categorized as Class B, which are best extinguished by foam, carbon dioxide, or dry chemicals. Class C fires involve live electrical equipment, which must be extinguished by a nonconductive extinguishing agent, such as carbon dioxide or dry chemicals. Class D fires involve combustible metals such as magnesium or titanium, which must be extinguished by special dry chemical extinguishers.

CLASSIFICATION OF OCCUPANCY HAZARDS

The occupancy hazard rating is a way to classify an occupancy or a building to determine the extent of sprinkler system requirements.

According to NFPA, there are three categories of hazard occupancies:

- **Light hazard:** Occupancies where the quantity and combustibility of contents are low. Fires in this category of occupancies tend to develop at a relatively low rate with low rates of heat release. Light hazard occupancies are typically institutional, educational, religious, residential, and commercial properties.
- **Ordinary hazard:** Ordinary hazard occupancies are divided into two types: Group 1 and Group 2.
  Group 1 includes occupancies consisting of materials low in combustibility, moderate quantity of combustibles, and stockpiles of combustibles not exceeding 8 feet. Occupancies in this group include automobile showrooms, bakeries, canneries, electronic plants, laundries, parking garages, and restaurant serving areas.
  Group 2 includes occupancies where quantity and combustibility of materials is moderate to high, with the stockpiles of combustibles not exceeding 12 feet. Examples of occupancies in this group include cereal mills, confectioners, distilleries, feed mills, libraries, machine shops, paper mills, and wood product assembly plants.
- **Extra hazard:** Occupancies where the quantity and combustibility of materials is very high are classified as extra hazard. Fires in this category of occupancies develop rapidly with a high rate of heat release. Extra hazard occupancies are also divided into two types: Group 1 and Group 2.
  Group 1 includes occupancies having hydraulic systems with flammable or combustible hydraulic fluids under pressure. Properties with process machinery that use flammable or combustible liquids in closed systems and those

having dust and lint in suspension are also included in this group. Aircraft hangars, plywood manufacturing plants, rubber vulcanizing plants, and sawmills belong to this group.

Group 2 type contains larger amounts of flammable or combustible liquids than Group 1 type. Examples of occupancies in this group include asphalt saturating, open oil quenching, plastics processing, and varnish and paint dipping.

METHODS FOR FIRE DETECTION

The National Fire Protection Association has developed NFPA No. *74*, Standard for Household Fire Warning Equipment in order to protect the customers and establish criteria for the manufacture and installation of residential fire detection systems.

There are two types of residential fire detection systems: smoke detectors and heat detectors. The basic residential detection system relies primarily on the use of smoke detectors.

Flame detectors are used to detect the direct radiation of a fire. They are mainly used in industrial processes and for the protection of combustion equipment.

Smoke Detectors There are two kinds of smoke detectors: ionization and photoelectric. The **ionization smoke detector** operates on the principle of changing the conductivity of air. It uses a radioactive source to ionize the air between two charged surfaces. The ionization of air causes a small flow of electrical current. When smoke from a fire enters the detection chamber, its presence causes a reduction in the current's flow. The electronic circuitry senses the reduced flow and triggers an alarm. An ionization detector is capable of sensing microscopic particles of combustion and is the best for detecting fire at an incipient stage.

A **photoelectric smoke detector** operates on the principle of the scattering of light. It consists of an LED (light emitting diode) light source, a supervisory photocell directly opposite the light source, and an alarm photocell within the detection chamber. When smoke is present in the chamber, the alarm photocell, which is located at right angles to the light beam, senses the light scattered off the smoke particles. This activates the electric circuit to activate an alarm. This device is the best for detecting fire at smoldering stage, when smoke particles are visible to the naked eye.

Heat Detectors The NFPA standard does not require the use of heat detectors as part of the basic protection scheme, but it recommends that heat detectors be used to supplement the basic smoke detector system. There are two types of heat detectors: fixed temperature and rate-of-rise.

Fixed temperature heat detectors may either be self-restoring or nonrestoring type. The self-restoring type consists of an open contact held by a bimetallic element. When the ambient temperature reaches a fixed setting (usually 135°F or 185°F), the contact is closed, thereby triggering an alarm. The contact will return to open position when the ambient temperature returns to normal. The nonrestoring (i.e., nonresettable) type utilizes a special alloy retainer designed to melt at a specific temperature. When the ambient temperature reaches the fixed setting, the fusible alloy melts to close the electrical contacts and to initiate an alarm.

A **rate-of-rise type heat detector** reacts when the temperature in the immediate vicinity rises higher than the preset rate per unit of time. It incorporates a sealed but slightly vented air chamber and a flexible metal diaphragm. Air within the chamber expands and contracts due to the fluctuation of temperature under normal conditions. The vent, which is calibrated, releases or admits air to compensate for the changes in pressure. When the temperature within the vicinity of the device rises quickly, air within the chamber expands faster than it can be vented. The resulting pressure distends the diaphragm and closes a set of normally open contacts, thereby activating an electrical circuit.

Flame Detectors Flame detectors are of two types: infrared and ultraviolet. Infrared radiation is present in most flames. Therefore, an **infrared (IR)** detector can detect the presence of fire instantaneously. A relatively long IR wavelength allows it to penetrate smoke, making detection possible. Because there are many sources of IR other than fire in an industrial setting, it is possible that a simple IR detector may cause false alarms.

Ultraviolet (UV) flame detectors operate by detecting UV radiation produced by fire. These detectors are not plagued by as many possible sources of false alarms. Sources of false alarms for UV detectors are well defined. The most common are lightning, X rays, and arc welding.

Dual spectrum flame detectors are also available that employ UV sensors in combination with a narrowband IR detector. To prevent false alarms caused by nonfire sources, the UV-IR detector utilizes an IR detector sensitive to wavelengths in the range of 4.1 to 4.6 microns. All hydrocarbon fires produce radiation within this range.

METHODS OF FIRE CONTROL

Fire has a need for oxygen, fuel, and heat. When deprived of any of these needs, there will be no fire. The most common method of fire suppression is the use of water. A universal medium for extinguishing fire, water is available in large quantities and is less expensive than any other fire-suppression medium. It works as a cooling agent by lowering the ignition temperature of combustion and also deprives the source of combustion from oxygen supply. Water supply for fire suppression may be provided by either manual or automatic systems. The manual systems use standpipes and hose, and the automatic systems use sprinklers.

Automatic Sprinkler Systems Automatic sprinkler systems consist of a network of pipes connecting water supply to a series of sprinkler heads. They provide automatic fire suppression in areas of a building where fire temperature has reached a predetermined level. Automatic sprinkler systems are the most effective in the suppression of Class A fire that contains wood, paper, and plastics.

The sprinklers are generally installed in a gridiron pattern near the ceiling. There are four basic types of sprinkler systems: wet pipe, dry pipe, preaction, and deluge.

Wet pipe type is the most common sprinkler system. In this system, the piping network contains water at all times. Water is immediately discharged onto the fire when a sprinkler activates (Figure 5.2).

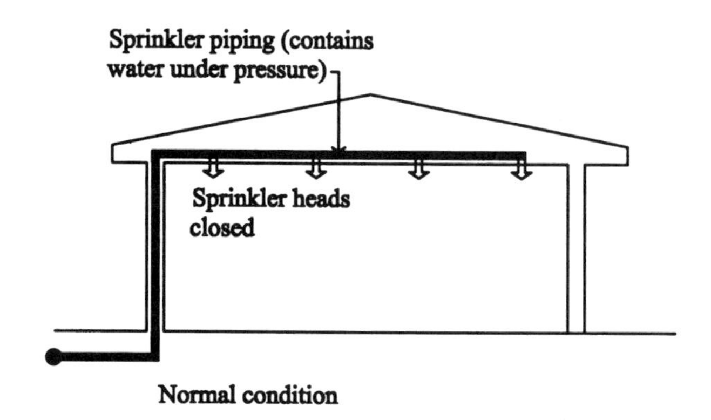

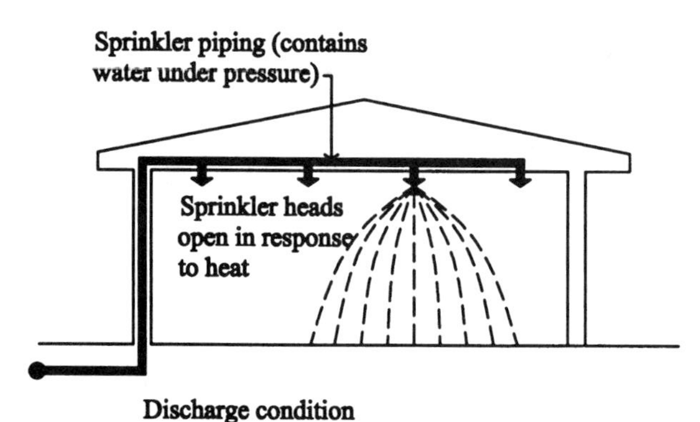

FIGURE 5.2 Wet pipe system

This system is very simple, reliable, and relatively inexpensive to install and maintain. The main disadvantage of the system is that it is not suitable for subfreezing environments.

A **dry pipe** sprinkler piping system is filled with pressurized air or nitrogen (Figure 5.3). The system is connected to a dry pipe valve that remains closed under normal conditions. When fire causes one or more sprinklers to operate, air within the system escapes and the dry pipe valve is released. Water is thus allowed to enter the piping network and flow through the open sprinkler heads. The main advantage of this system is its capability to provide protection to spaces that are subject to freezing temperature conditions. It is more complex than a wet pipe system and requires greater attention in design, installation, and maintenance. It may produce greater fire damages because of a higher response time. A dry pipe system must be thoroughly drained and dried following operation in order to prevent the corrosion of pipes.

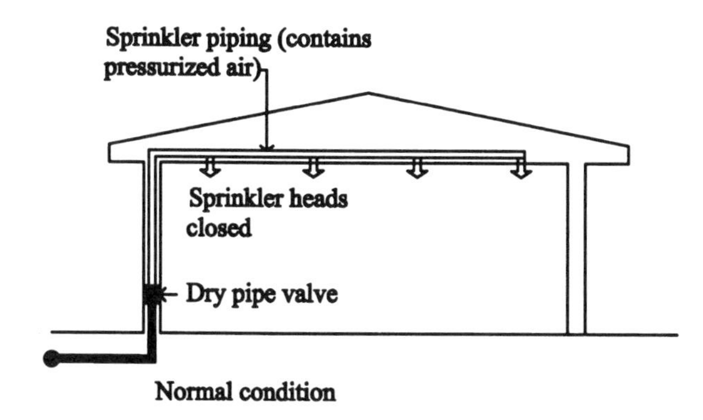

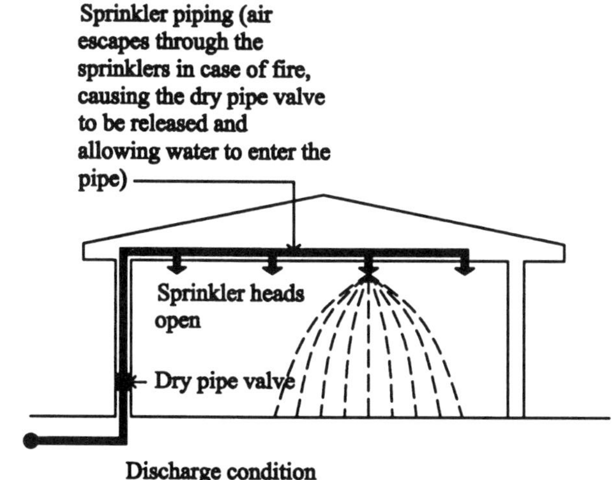

FIGURE 5.3 Dry pipe system

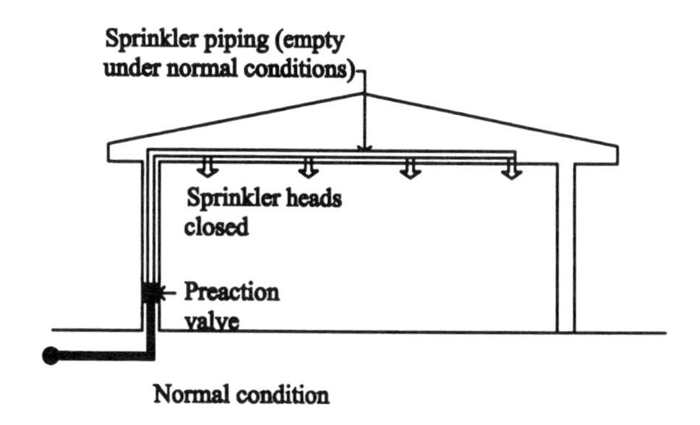

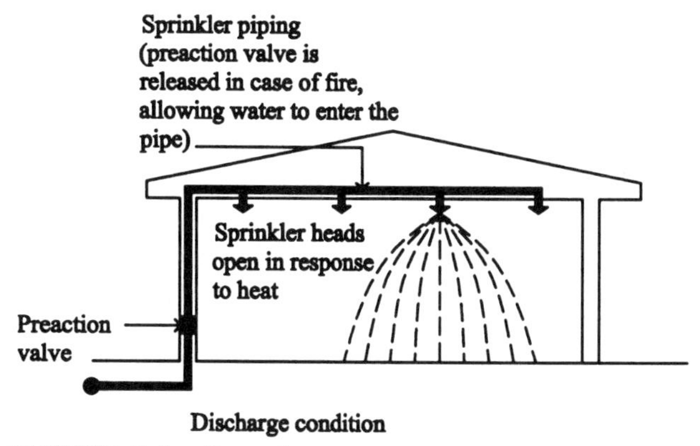

FIGURE 5.4 Preaction system

A **preaction** system employs the basic concept of a dry pipe system. It consists of a preaction valve that controls the flow of water in the sprinkler piping network (Figure 5.4). Discharge in the system is initiated by two independent events. When the detection device identifies a developing fire, the preaction valve is opened, allowing water to enter the pipes. The sprinklers also open independently in response to the heat from fire and discharge water. A preaction system with double interlock allows water to enter the sprinkler piping when both the detection device and the sprinklers identify a developing fire. This system provides an added level of protection against any inadvertent discharge and is suitable for areas where water damage is a serious problem.

A **deluge** system is a variation of preaction, equipped with open-type sprinklers (Figure 5.5). Actuation of the fire detection systems releases a control valve, which in turn causes immediate water flow through all sprinklers installed in the network. Since all sprinkler heads are open, every sprinkler on a deluge system will discharge water simultaneously when the control valve is released. This system is recommended for buildings where the spread of fire is anticipated to be rapid and for very high hazard areas.

Standpipe Systems A standpipe system is required for high-rise buildings where the hose from fire-fighting equipment cannot reach the upper floors. It provides fire hose stations for manual application of water to fires in buildings. The system consists of piping, valves, hose racks, hose connections, and auxiliary equipment necessary to provide sprays of water to suppress fire. Standpipes are classified into three categories.

- **Class I:** This category of standpipe supplies 2½″ hose outlets at each floor level for use by the fire department. The 2½″ outlet hose valves furnish the firefighters with a water hydrant

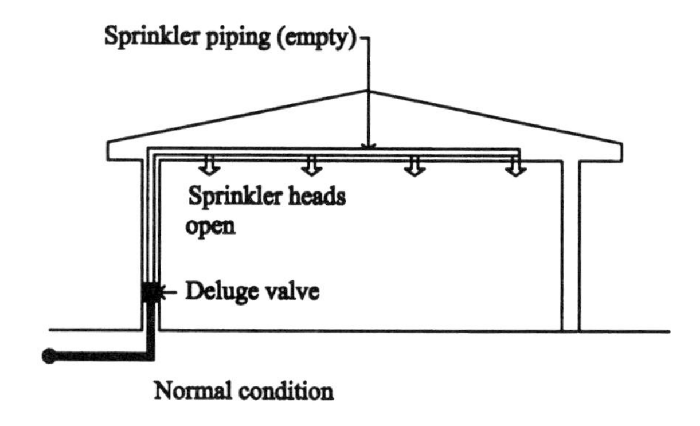

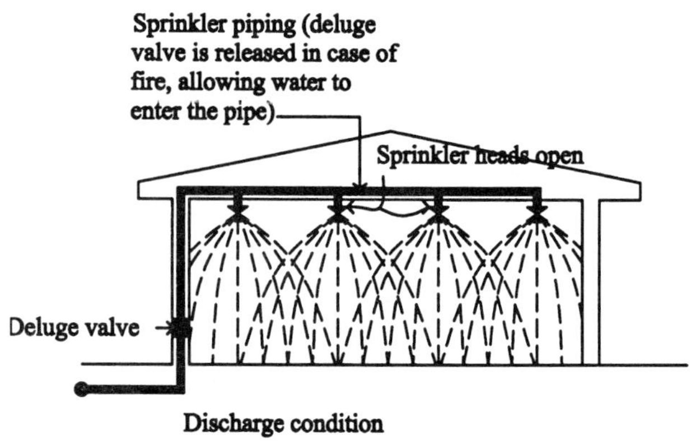

FIGURE 5.5 Deluge system

and adequate water supply for using hose stream during fire. The minimum water supply requirement for Class I category is 500 gpm for the first standpipe and 250 gpm for each additional standpipe.

- **Class II:** This category of standpipe is designed to be used by the occupants of the building until the arrival of the firefighters. It supplies 1½″ hose stations along with a hose rack. The minimum water supply requirement for this category is 100 gpm. Hose stations should be located so that all spaces on a floor are within 30 feet of a nozzle attached to a 100-foot-long hose.
- **Class III:** This is a combination of Class I and Class II standpipes. It supplies both 2½″ hose outlets and 1½″ hose stations.

A fire department pumper connection, known as Siamese connection, is required to be provided outside the building. It is a two-way connection having two 2½″ outlets supplied by a 4″ pipe.

Maximum height for a standpipe system is 275 feet. For buildings higher than this limit, separate standpipe systems are required for each additional 275 feet or less.

The water pressure requirement for all categories of standpipes is a supply that will provide 65 psi of residual pressure at the highest outlet on the standpipe. Residual pressure is the pressure reading on the gauge when the required quantity of water is flowing through the hose.

Water supply for standpipes may be supplied from city water main, gravity tank, or pressure tank. It must be ensured that the capacity of the source is adequate to furnish the total demand for at least 30 minutes.

The standpipes may either be wet or dry types. The pipes in a wet system are always filled with water under pressure. Water from such a system will flow through the hose as soon as it is activated.

Pipes in a dry system are generally filled with compressed air instead of water. A valve has to be operated in order to admit water into the system. Dry pipe is used in areas that are subject to freezing conditions.

DESIGN OF AUTOMATIC SPRINKLER SYSTEMS

Piping Design Piping for an automatic sprinkler system can be designed using a **hydraulic method.** The pipe sizes can be determined using the following formula:

$$p_f = (4.52 \cdot Q^{1.85})/(C^{1.85} \cdot \partial^{4.87}) \qquad (1)$$

where

p_f = frictional resistance, in psi per foot of pipe
Q = flow rate, in gpm
C = friction loss coefficient of piping material (100 for cast iron; 120 for black steel, galvanized steel, and plastic;140 for cement-lined cast iron; and 150 for copper and stainless steel)
∂ = internal diameter of pipe in inches

The orifice size of a sprinkler and the residual pressure of the water supply affect the **flow rate** of a sprinkler in any sprinkler system. It is calculated using the following formula:

TABLE 5.1

Pipe Schedule for Number of Sprinklers Allowed in an Automatic Sprinkler System

Pipe Size	*Hazard Classification*					
	Light		*Ordinary*		*Extra*	
	Steel	*Copper*	*Steel*	*Copper*	*Steel*	*Copper*
1″	2	2	2	2	1	1
1¼″	3	3	3	3	2	2
1½″	5	5	5	5	5	5
2″	10	12	10	12	8	8
2½″	30	40	20	25	15	20
3	60	65	40	45	27	30
3½″	100	115	65	75	40	45
4	a	a	100	115	55	65
5			160	180	90	100
6			275	300	150	170
8			b	b		

[a]One 4-inch system may serve up to 52,000 sq. ft. of floor area.
[b]One 8-inch system may serve up to 52,000 sq. ft. of floor area.

$$Q = K \cdot \sqrt{p} \qquad (2)$$

where

Q = flow rate in gpm
K = flow constant, per unit; varies from 1.3 to 1.5 for a ¼″ orifice, 5.3 to 5.8 for a ½″ orifice, and 13.5 to 14.5 for a ¾″ orifice
p = residual pressure in psi

A simple method of piping design for small automatic sprinkler systems is the use of **piping schedule.** The sizes of branch pipes and risers are determined from a pipe schedule (see Table 5.1), assuming that residual pressure and flow rate are in compliance with code. The method is very useful for preliminary design and cost estimates.

Sprinklers On a Branch Line A maximum of 8 sprinklers are allowed to be installed on a branch line, on either side of a cross main of a sprinkler system designed for either a light or an ordinary hazard occupancy. The limit for extra hazard occupancy is 6 sprinklers.

Protection Area The maximum floor area that can be protected by a single sprinkler system shall not exceed 52,000 square feet for light or ordinary hazards. The maximum area protected by a single system for extra hazard occupancy shall not exceed 25,000 square feet if the piping is designed using the pipe schedule method, and 40,000 square feet if it is designed using the hydraulic method.

The actual number of sprinklers to be installed and their spacing depend on the hazard classification of the building. Table 5.2 gives the maximum sprinkler protection area and spacing between sprinklers.

WATER DEMAND FOR FIRE PROTECTION

A small part of the sprinkler system needs to be operated during the early stage of a fire. Therefore, fire protection codes require only a small area of the building to be taken into consideration for simultaneous water demand. This space is called the area of sprinkler operation. The area selected should be the one that is the most remote from the source of water supply. The size of the area of sprinkler operation varies from 1,500 square feet for light hazard to 5,000 square feet for extra hazard occupancies. A minimum density of water flow has to be maintained

TABLE 5.2

Maximum Spacing Between Sprinklers and Coverage Per Sprinkler

Construction Type		*Classification of Hazard*		
		Light	*Ordinary*	*Extra*
Unobstructed	Spacing (ft.)	15	15	12
	Area (sq. ft.)	200[a]–225[b]	130	90[a]–130[b]
Obstructed, noncombustible	Spacing (ft.)	15	15	12
	Area (sq. ft.)	200[a]–225[b]	130	90[a]–130[b]
Obstructed, combustible	Spacing (ft.)	15	15	12
	Area (sq. ft.)	130[c]–225[d]	130	90[a]–130[b]

[a]To be used when pipe sizing is based on pipe schedule.
[b]To be used when pipe sizing is based on hydraulic method.
[c]To be used when framing members are spaced less than 3 feet on center.
[d]To be used when framing members are spaced 3 feet or more on center.

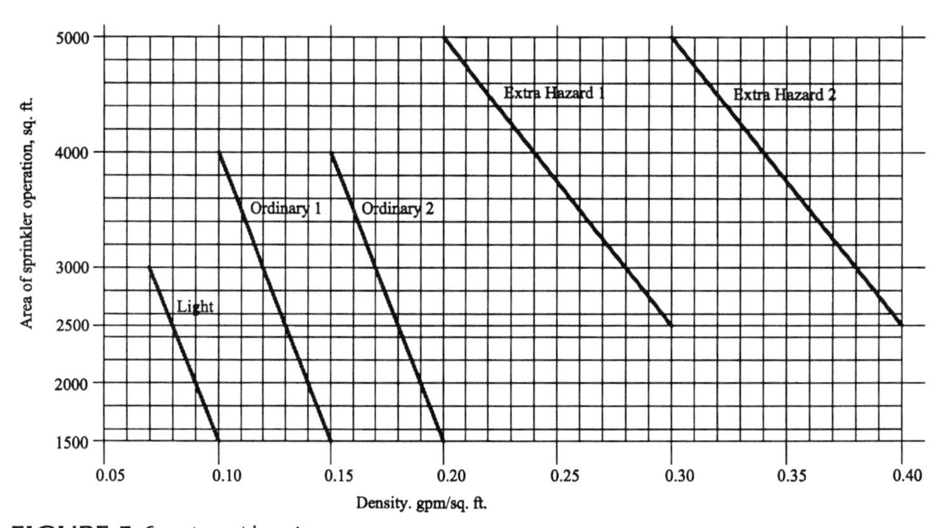

FIGURE 5.6 Area/density curves

for different areas of sprinkler operation. The relationships between area of operation and water flow density is shown in Figure 5.6.

The minimum water requirement for fire protection is determined independently of the actual size of a building. This is because a fire can usually be expected to start in only one area. Therefore, total water demand is calculated using the area of sprinkler operation. Water demand of hoses is also included if the sprinkler systems are supported by standpipe systems. Table 5.3 shows water demand by hose pipes and the required duration of the supply. The total water demand for combined sprinkler and standpipe systems can be calculated using the following formula:

$$\text{TWD} = (\text{ASOP} \cdot D \cdot \text{OVF}) + \text{HSD} \qquad (3)$$

where

TWD = total water demand in gpm
ASOP = area of sprinkler operation in sq. ft.
D = density of water flow in gpm/sq. ft.
OVF = overage factor (usually 1.1)
HSD = hose stream demand in gpm

Supply pipe sizes for standpipe connections considering the total water demand are shown in Table 5.4.

TABLE 5.3

Water Requirements by Hose Pipes and Duration of Supply

Hazard Type	*Inside Hose (gpm)*	*Inside + Outside Hose (gpm)*	*Duration (minutes)*
Light	0, 50, or 100	100	30
Ordinary	0, 50, or 100	250	60–90
Extra	0, 50, or 100	500	90–120

TABLE 5.4

Supply Pipe Sizes for Standpipe Connections

Total Accumulated Flow in GPM	*Total Distance of Piping From Furthest Outlet*		
	Less Than 50 Feet	*50–100 Feet*	*Over 100 Feet*
100	2″	2½″	3″
101–500	4″	4″	6″
501–750	5″	5″	6″
751–1250	6″	6″	6″
Over 1250	8″	8″	8″

OTHER METHODS OF FIRE SUPPRESSION

Alternative methods of fire suppression are available in situations where the use of water may not be a practical solution. These methods include the use of carbon dioxide, dry chemicals, foam systems, and halogenated gas.

Carbon Dioxide Carbon dioxide is a colorless and odorless gas. It can extinguish a fire without leaving any residue. The introduction of CO_2 into a fire displaces oxygen in the atmosphere thereby retarding the combustion process. At the same time, the extinguishing process is aided by a reduction in the concentration of gasified fuel in the fire area. CO_2 also provides some cooling in the combustion zone, completing the extinguishing process. The evaporation of the liquefied gas absorbs about 120 BTU (British thermal unit) of heat per pound of the gas.

CO_2 is effective in confined and unventilated spaces. Such spaces include display areas, electrical and mechanical chases, and unventilated areas above suspended ceilings that are not inhabited by people or other living creatures.

Dry Chemicals The best candidates for dry chemicals that suppress fire are industrial applications. They are extremely effective fire-fighting agents that suppress fire by covering the surface of the fuel. The coating separates the fuel from the oxygen supply, and thus retards fire. The dry chemicals usually contain bicarbonates, chlorides, and phosphates.

Foam Systems Foams are gas-filled bubbles, produced by using a generator that mixes water with detergent or other chemicals. It is possible to produce about 1,000 gallons of foam using one gallon of water. Foam systems are most effective for fires involving flammable liquids. Being lighter than the combustible liquids, they float on the surfaces of the liquids, thereby insulating them from oxygen. Moreover, the vaporization of water contained in the films of the foam provide a certain measure of cooling by heat absorption. A good foam blanket strongly resists the heat and flame of a fire.

Halogenated Agents Halogenated agents are gases that contain halogen atoms (bromine, chlorine, fluorine, or iodine). The most commonly used halogenated gas is Halon 1301. It is a very effective fire suppressant that extinguishes fire by inhibiting the chemical reaction of fuel and oxygen. But scientific evidence indicates that the agent is an ozone-depleting chemical. Therefore, the use of halogenated agents has been banned; they are being replaced with a number of alternatives such as IG-541 (commonly known as INERGEN) and HFC-227 (commonly known as FM-200). These agents have either zero or negligible ozone depleting potential. Their effectiveness as fire suppressants is still being evaluated.

5.1 WATER FOR HVAC

WATER AS A MEDIUM FOR HEATING AND COOLING

Water is a chemically stable, nontoxic, and inexpensive fluid. It also has a remarkable capability of carrying large quantities of heat. Compared to air, it is a much less bulky medium for conveying heat or cooling. One pound of air is about 14 cubic feet and can carry 0.24 BTU of heat per 1°F temperature difference; one pound of water, on the other hand, has volume of only about 0.016 cubic foot and is capable of carrying 1 BTU of heat per 1°F temperature difference. Therefore, a system of water piping takes up much less space of building structure than ductwork for air required for an equivalent amount of thermal distribution. Water is, therefore, a very popular fluid used in HVAC systems, particularly for large buildings.

Water can be used for both heating and cooling. For heating purposes, the temperature of water is raised to 160°F to 250°F or more in a boiler. It is lowered to 40°F to 50°F in a chiller for cooling purposes. Pipes are used to convey the product from the boiler or chiller to an air-handling unit, which transfers the heat or cooling to an airstream for ducted delivery to different spaces in the building. In small buildings, heated or chilled water is piped to a terminal device that transfers the heat or cooling directly to the room air.

WATER DEMAND FOR HEATING AND COOLING

The water demand for heating or cooling is dependent on the total quantity of heat required to be added to or removed from a building. It also depends on the designed temperature difference of the flow of water entering and leaving the heat exchanger. In other words, it is the function of the peak heating or cooling load of a building and the required rise or drop in temperature of water used in the process.

Flow rate or demand of water is calculated in terms of gallons per minute (gpm). One gallon per minute equals 60 gallons per hour. Since a gallon of water weighs 8.33 pounds, one gallon per minute equals $8.33 \times 60 \approx 500$ pounds per hour. Using this relationship, water demand for heating or cooling can be calculated as follows:

$$\mathrm{FL} = Q/500 \cdot (t_1 - t_2) \qquad (4)$$

where

- FL = flow rate of water, in gpm
- Q = heat required to be added or removed, in BTUH (BTU per hour)
- $t_1 - t_2$ = difference between supply and return water temperatures

For heating systems, a temperature drop of 20°F between supply and return water temperatures is an industry standard. Therefore, heat transferred by 1 gpm of water equals $500 \cdot 20 = 10{,}000$ BTUH. A building with a heating load of 50,000 BTUH will require 5 gpm of water at 20°F drop in temperature between supply and return.

PIPING

Many types of materials are used for HVAC piping. Most common are Schedule 40 steel pipe, copper pipe (types K, L, and M), and Schedule 80 CPVC (thermoplastic) pipe. Pipe size is based on either

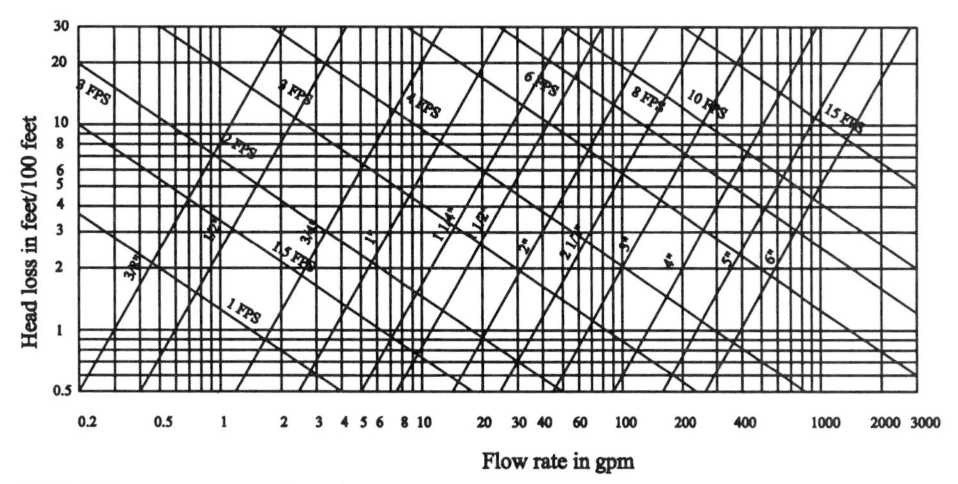

FIGURE 5.7 Sizing chart for copper pipes

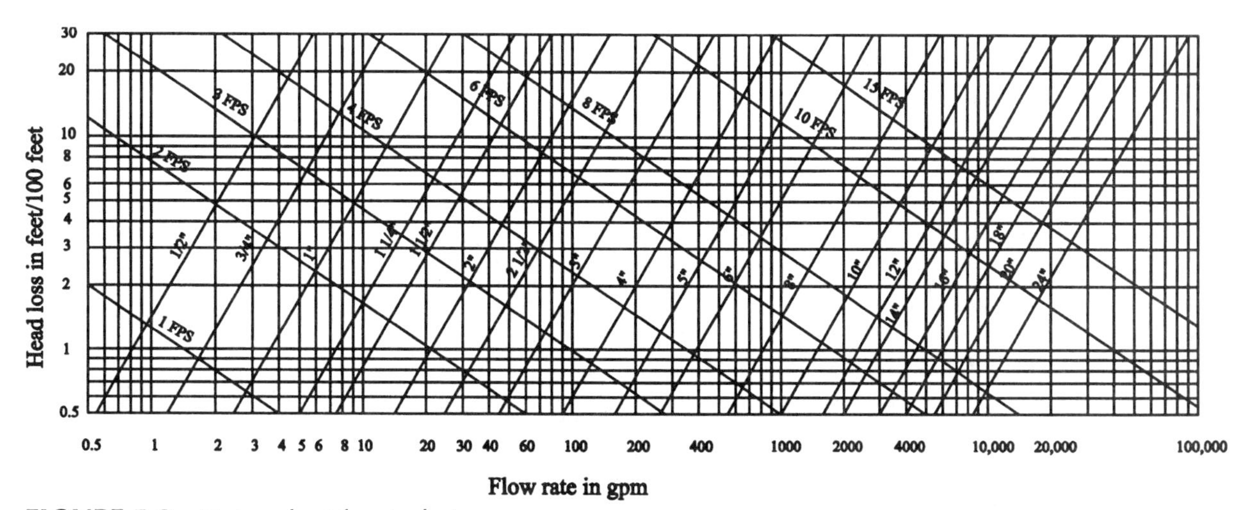

FIGURE 5.8 Sizing chart for steel pipes

velocity or head loss, both of which provide roughly equal results. Head or friction loss is the reduction of water pressure due to the friction that the walls of the pipe impose on a liquid—i.e., a measure of the resistance of the piping system to the flow of the water through it. It is dependent on the viscosity of the fluid and the turbulence of the flow. Head or friction loss in feet of water is equal to 2.31 feet/psi. A friction loss of 2.5 feet per 100 feet of pipe length is used as design criterion.

Pipe sizes should be large enough to minimize the noise produced by water flow. Larger pipe sizes also help in reducing the amount of friction that the pump has to overcome. Water velocities of 3 to 5 feet per second are generally recommended for piping design. Once the flow rate or demand of water has been established, the size of HVAC pipes can be determined using sizing charts (Figures 5.7 and 5.8).

Water-bearing HVAC pipes have to be insulated to minimize the transfer of heat to the surrounding unconditioned air before the water reaches the terminal device. Insulation can increase the pipe diameter by 2 to 8 inches. Added precaution has to be taken in insulating chilled water pipes to prevent condensation on the pipe surfaces.

HVAC WATER QUALITY

Water quality is important for both heating and cooling purposes in order to maintain the reliability and efficiency of the equipment, such as boilers, chillers, and cooling towers. Mineral scale deposit, corrosion, and growth of microorganisms may cause failure of the systems.

Boiler Water Scale-forming minerals such as calcium and magnesium, introduced by makeup water, primarily cause scale and sludge deposit in a boiler system. The primary means for controlling boiler scale caused by minerals in the makeup water is to treat the water, using ion exchange softening or demineralization. This treatment should remove the majority of the scale-causing minerals.

Oxygen that enters a boiler system may contribute to accelerated corrosion of the feedwater system, boiler, and condensate return system. De-aeration of makeup water is helpful in controlling this type of corrosion. Once the makeup water is de-aerated, various organic and inorganic oxygen-scavenger (40 to 60 mg/L of sulfite, for example) compounds are added to complete oxygen removal. Oxygen scavengers should be added to the boiler makeup water stream immediately after de-aeration so as to protect all downstream components of the system.

Inner surfaces of a boiler system must also be protected against corrosion caused by contact with water. The best method of protection is to simply maintain the boiler water at a pH-value between 8 and 8.5. This will make the water noncorrosive to steel boiler internals. It is also required to control carbonate and phosphate scale formation that occurs at higher pH-value levels.

Despite the treatment of makeup water, some dissolved minerals may enter the boiler system. This material remains in the drum as solids. Excessive concentration of these solids will eventually lead to unsatisfactory operation of the system. Boiler blowdown is necessary under such circumstances. It means a complete or partial removal of boiler water and replacing it with treated water.

Chiller and Cooling Tower Water Water for cooling is also required to be treated to remove dissolved solids and to prevent scale formation. Growth of microorganisms has to be controlled to prevent tube fouling. Chlorine and ozone work well on microorganisms. Since the chilled water loop is usually a closed one, it needs to be treated as often as makeup water is added.

Open cooling tower systems are subject to deposits of airborne dust and debris. Fine particles of these foulants tend to collect in the condenser-water system, get deposited on the heat-transfer surfaces in the form of sticky mud, and interfere with efficient operation of the system. These suspended solids are removed by filtration using strainers and filters on cooling tower systems. Makeup water for cooling systems, containing turbidity or suspended matter, should be treated before use by coagulation and filtration.

Microbiological growths such as algae, slime-forming bacteria, and mold take place in cooling towers. Algae and fungi can cause serious problems by blocking the air in cooling towers and plugging the distribution systems. Deposits of bacterial slime on heat-transfer surfaces will affect the efficiency of the cooling system. Traditional biocides, such as gaseous or liquid chlorine, are quite effective for removal of these organic growths. Use of ozone is also a common practice for treatment of cooling tower water.

Cooling tower blowdown is periodically done in order to control the concentration of dissolved solids by systematic drainage of a portion of the circulating water. Water thus removed from the system has to be replaced with treated water.

REVIEW QUESTIONS

1. What are the basic factors of fire?
2. What category of fire involves flammable liquids?
3. What type of detector is the most suitable for detection of fire at a very early stage?
4. Name an automatic fire sprinkler system that is suitable for protection of spaces subject to freezing temperature conditions.
5. What is the maximum number of sprinklers recommended to be installed on branch pipe?
6. Why have halogenated agents been banned as fire suppressants?
7. How much heat can be carried by a gallon of water when temperature drop is 1°F?
8. What is the recommended velocity of water in HVAC piping?
9. Name the minerals, the presence of which in boiler water will cause scale formation in the system.
10. What is the recommended pH-value of water for a boiler system

ANSWERS

1. Fuel, heat, and oxygen
2. Class B
3. Ionization smoke detector
4. Dry pipe system
5. 8 sprinklers
6. Halogenated agents have high ozone-depleting potential
7. 1 gallon = 8.33 lbs.; Heat carried by 1 lb. of water @ 1°F temperature drop = 1 BTU; Heat carried by 8.33 lbs. of water @ 1°F temperature drop = 8.33 BTU.
8. 3 to 5 feet per second
9. Calcium and magnesium.
10. 8 to 8.5.

INDEX

Tables and Maps